Springer Theses

Recognizing Outstanding Ph.D. Research

Aims and Scope

The series "Springer Theses" brings together a selection of the very best Ph.D. theses from around the world and across the physical sciences. Nominated and endorsed by two recognized specialists, each published volume has been selected for its scientific excellence and the high impact of its contents for the pertinent field of research. For greater accessibility to non-specialists, the published versions include an extended introduction, as well as a foreword by the student's supervisor explaining the special relevance of the work for the field. As a whole, the series will provide a valuable resource both for newcomers to the research fields described, and for other scientists seeking detailed background information on special questions. Finally, it provides an accredited documentation of the valuable contributions made by today's younger generation of scientists.

Theses are accepted into the series by invited nomination only and must fulfill all of the following criteria

- They must be written in good English.
- The topic should fall within the confines of Chemistry, Physics, Earth Sciences, Engineering and related interdisciplinary fields such as Materials, Nanoscience, Chemical Engineering, Complex Systems and Biophysics.
- The work reported in the thesis must represent a significant scientific advance.
- If the thesis includes previously published material, permission to reproduce this must be gained from the respective copyright holder.
- They must have been examined and passed during the 12 months prior to nomination.
- Each thesis should include a foreword by the supervisor outlining the significance of its content.
- The theses should have a clearly defined structure including an introduction accessible to scientists not expert in that particular field.

More information about this series athttp://www.springer.com/series/8790

Stefan Thiele

Read-Out and Coherent Manipulation of an Isolated Nuclear Spin

Using a Single-Molecule Magnet
Spin-Transistor

Doctoral Thesis accepted by
the University of Grenoble, France

 Springer

Author
Dr. Stefan Thiele
Sensirion AG
Staefa
Switzerland

Supervisor
Dr. Wolfgang Wernsdorfer
Nano, Institut Neel
Grenoble
France

ISSN 2190-5053   ISSN 2190-5061   (electronic)
Springer Theses
ISBN 978-3-319-79574-4   ISBN 978-3-319-24058-9   (eBook)
DOI 10.1007/978-3-319-24058-9

Springer Cham Heidelberg New York Dordrecht London
© Springer International Publishing Switzerland 2016
Softcover re-print of the Hardcover 1st edition 2016

Printed on acid-free paper

Springer International Publishing AG Switzerland is part of Springer Science+Business Media
(www.springer.com)

Supervisor's Foreword

It is with great pleasure and the highest enthusiasm that I write this foreword to Stefan Thiele's thesis, which reports his most exciting results. Its endeavor is driven by one of the most ambitious, technological goals of today's scientists: the realization of an operational quantum computer. In this regard, the basic building block is generally composed of a two-level quantum system, namely a quantum bit (or qubit). Such a quantum system must be fully controllable and measurable, which requires a connection to the macroscopic world. In this context, solid-state devices, which establish electrical interconnections to the qubit, are of high interest, mainly due to the variety of methods available for fabrication of complex and scalable architectures. Moreover, outstanding improvements in the control of the qubit dynamics have been achieved in the last years. Among the different solid-state concepts, spin-based devices are very attractive because they already exhibit relatively long coherence times. For this reason, electrons possessing a spin 1/2 are conventionally thought as the natural carriers of quantum information. However, the strong coupling to the environment makes it extremely difficult to maintain a stable entanglement. Alternative concepts propose the use of nuclear spins as building blocks for quantum computing, as they benefit from longer coherence times compared to electronic spins, because of a better isolation from the environment. But weak coupling comes at a price: the detection and manipulation of individual nuclear spins remain difficult tasks. In this context, the main objective of the Ph.D. of Stefan Thiele was to lay the foundation of a new field called molecular quantum spintronics, which combines the disciplines of spintronics, molecular electronics, and quantum information processing. In particular, the objective was to fabricate, characterize, and study molecular devices (molecular spin-transistor, molecular spin-valve and spin filter, molecular double-dot devices, carbon nanotube, nano-SQUIDs, etc.) in order to read and manipulate the spin states of the molecule and to perform basic quantum operations. The visionary concept of the project is underpinned by worldwide research on molecular magnetism and supramolecular chemistry, and in particular within the European Institute of Molecular Magnetism (http://www.eimm.eu/), and collaboration with outstanding scientists in the close environment. For the project, we found the following funding: the main contributions came from the French Research Agency (ANR), the Contrat Plan Etat Région (CPER), the European Research Council (ERC), the Réseau Thématique de Recherche Avancé (RTRA), and the support from our institute. The Ph.D. of Stefan Thiele was funded by an ERC advanced grant. Among the most important results before Stefan Thiele's Ph.D., we showed the possibility of magnetic molecules to act as building blocks for the design of quantum spintronic devices and demonstrated the first important results in this new research area. For example, we have built a novel spin-valve device in which a nonmagnetic molecular quantum dot, consisting of a single-wall carbon nanotube contacted with nonmagnetic electrodes, is laterally coupled via supramolecular interactions to a $TbPc_2$ molecular magnet [Ph.D. of Matias Urdampilleta (2012)]. The localized magnetic moment of the SMM led to a magnetic field-dependent modulation of the conductance in the nanotube with magnetoresistance ratios of up to 300 % at low temperatures. We also provided the first experimental evidence for a strong spin–phonon coupling between a single-molecule spin and a carbon nanotube resonator [Ph.D. of Marc Ganzhorn (2013)]. Using a molecular spin-transistor, we achieved the electronic read-out of the nuclear spin of an individual metal atom embedded in a single-molecule magnet (SMM) [Ph.D. of Romain Vincent (2012)]. We could show very long spin lifetimes (several tens of seconds). Here, the Ph.D. of Stefan Thiele started with a completely new breakthrough. He proposed and demonstrated the possibility to perform quantum manipulation of a single nuclear spin by using an electrical field only. This has the advantage of reduced interferences with the device and less Joule heating of the sample. As an electric field is not able to interact with the spin directly, he used an intermediate quantum mechanical process, the so-called hyperfine Stark effect, to transform the electric field into an effective magnetic field. His project was designed to play a role of pathfinder in this—still largely unexplored—field. The main target concerned fundamental science, but applications in quantum electronics are expected in the long run.

Grenoble   Dr. Wolfgang Wernsdorfer
April 2015   Research Director

Abstract

The realization of a functional quantum computer is one of the most ambitious, technological goals of today's scientists. Its basic building block is composed of a two-level quantum system, namely a quantum bit (or qubit). Among the other existing concepts, spin-based devices are very attractive because they benefit from the steady progress in nanofabrication and allow for the electrical read-out of the qubit state. In this context, nuclear spin-based devices exhibit additional gain of coherence time with respect to electron spin-based devices due to their better isolation from the environment. But weak coupling comes at a price: the detection and manipulation of individual nuclear spins remain challenging tasks.

Very good experimental conditions were important for the success of this project. Besides innovative radio frequency filter systems and very low noise amplifiers, I developed new chip carriers and compact vector magnets with the support of the engineering departments at the institute. Each part was optimized in order to improve the overall performance of the setup and evaluated in a quantitative manner.

The device itself, a nuclear spin qubit transistor, consisted of a $TbPc_2$ single-molecule magnet coupled to source, drain, and gate electrodes and enabled us to read out electrically the state of a single nuclear spin. Moreover, the process of measuring the spin did not alter or demolish its quantum state. Therefore, by sampling the spin states faster than the characteristic relaxation time, we could record the quantum trajectory of an isolated nuclear qubit. This experiment shed light on the relaxation time T_1 of the nuclear spin and its dominating relaxation mechanism.

The coherent manipulation of the nuclear spin was performed by means of external electric fields instead of a magnetic field. This original idea has several advantages. Besides a tremendous reduction of Joule heating, electric fields allow for fast switching and spatially confined spin control. However, to couple the spin to an electric field, an intermediate quantum mechanical process is required. Such a process is the hyperfine interaction, which, if modified by an electric field, is also referred to as the hyperfine Stark effect. Using the effect, we performed coherent rotations of the nuclear spin and determined the dephasing time T_2^*. Moreover, exploiting the static hyperfine Stark effect we were able to tune the nuclear qubit in and out of resonance by means of the gate voltage. This could be used to establish the control of entanglement between different nuclear qubits.

In summary, we demonstrated the first single-molecule magnet based quantum bit and thus extended the potential of molecular spintronics beyond classical data storage. The great versatility of magnetic molecules holds a lot of promises for a variety of future applications and, maybe one day, culminates in a molecular quantum computer.

Acknowledgements

First of all, I want to say thank you to all the people I encountered during my Ph.D. and that helped me to make a lot of very enjoyable memories and the time in Grenoble a very special part of my life. In particular, I want to thank my supervisor Wolfgang Wernsdorfer, who guided me through the stormy waters of the Ph.D. Thank you for all your support and for being available any time I needed your advice. I am very grateful that you let me join your team and most of all that you encouraged and taught me to pursue many ideas which went beyond my project. This aspect was broadening my horizon unlike anything else and made this thesis one of the most enriching experiences in my life. Last but not least, I want to thank you for taking care of my experiment every time I was in need. Moreover, I want to thank Franck Balestro. Thank you for letting me measure on your setup, providing me with samples, and all the He transfers you did for me. But, most of all, thank you for always finding the words to encourage me, both in my work and personal life and of course for helping me finding the sample of my Ph.D. I am also very grateful to you for reading this manuscript and reshaping it into its almost perfect state;-).

I also want to thank Markus Holzmann, who helped me understanding and performing quantum Monte Carlo simulations. I appreciate very much your enthusiasm and the cheerful working atmosphere. I want to thank Rafik Ballou, who did not give up until he found a theory to properly describe the hyperfine Stark effect. It was always a pleasure to discuss with you. I am also very thankful to my referees Jakob Reichel and Wulf Wulfhekel as well as to my thesis comitee members Vincent Jacques, Patrice Bertet, and Mairbek Chshiev for the sincere work and fruitful comments on my thesis. Moreover, I want to thank all the researchers at the Néel institut I was interacting with, especially Tristan Meunier, Olivier Buisson, and Nicolas Roch for their help and advice.

Furthermore, I want to thank all the present and former members of the NanoSpin group. Thank you Jean-Pierre, Vitto, and Oksana for being very enjoyable office mates and never being tired to discuss personal and work affairs. Many thanks to Edgar and Christophe, who helped me solving a lot of informatics problems. Thank you Jarno, Viet, Antoine, Raoul, and Matias for all the fruitful discussions and coffee breaks. Thank you Romain for letting me measure on your sample and introducing me to the glorious world of Python. Thank you Marc for all your help and support especially during the hard time after the worldcup. Thank you Clément, I am very glad you joined our group and it was always a pleasure to talk to you.

Merci à Eric Eyraud, qui m'a aidé énormément avec mon cryostat et tout ce qui vient avec. Sans toi rien n'était possible. Merci également pour m'avoir montré comment on peux faire le "produit". Je voudrais remercier aussi Daniel Lepoittevin qui m'avait beaucoup aidé avec tout ce qui concerne l'électronique de mesure et pour m'avoir expliqué tout son fonctionnement. Un grand merci également á Christophe Hoarau pour m'avoir aidé avec les mesures et simulations á haute fréquence. Merci beaucoup aussi á Yves Deschanels qui m'avait aidé avec la construction des bobines supra et qui a passé une semaine dans la cave avec moi. Un grand merci aussi à Didier Dufeu, Richard Haettel, David Barral et Laurent Del-Rey pour toute leur aide.

I want to thank also all my friends who I have not mentioned yet. Thank you Carina for convincing me to come to Grenoble and for all your help and support. Thank you Angela for sharing the burden of writing this thesis, and of course the secret tiramisu recipe. Thank you Liza for motivating us all to go out for beers and for lending me your appartment to celebrate my birthday. Thank you Peter for your honesty and very funny way to see things. Thank you Claudio and Elena for unbreakable good mood, even if Beatrice stole you the last bit of sleep. Thank you Sven for all parties we had at your place especially in the evening of my defense. Thank you Cornelia and Roman for nice skiing trips. Thank you Ovidiu, Martin, and Marc for the nice evenings at D'Enfert. Thank you Simone and Francesca for the enjoyable evenings with the Scarpe Mobile. Thank you Clément, Dipankar, Angelo, Christoph, Hanno, Farida, and Tobias for all the nice moments in and outside the lab. Merci également à Gosia, Julian, Marine, Juan Pablo, Elenore, Aude et Clément pour m'avoir reçu très chaleureusement dans votre groupe. Je voudrais remercier mon équipe de volley avec qui j'ai passé beaucoup des bonnes moments, en particulier Malo et Olivier pour la meilleur saison de volley de toute ma vie. Vielen Dank auch an meine Freunde aus der Heimat Lars, Christian, Stefan, Daniel, Toni, Anja, Micha, Stefan, Rico für die schönen Momente mit euch.

Je voudrais remercier ma copine Sophie, qui m'a beaucoup soutenu et encouragé pendant ces deux dernières années. Ta constante bonne humeur était toujours une grande motivation pour moi. Un grand merci également à toute ta famille et en particulier à Vincent, Claudine et Céline. Zum Schluss möchte ich ganz besonders meiner Familie danken, die mich die ganze Zeit nach Kräften unterstützt haben und ohne die all dies nicht möglich gewesen wäre. Ein ganz besonderer Dank gilt vor allem meinen Eltern Andreas und Ines, meiner Schwester Sabrina, und meinen Großeltern Hans und Inge, weil ihr immer für mich da gewesen seid, wenn ich euch brauchte.

Stefan Thiele

Contents

Chapter 1
Introduction

1.1 Molecular Spintronics

The computer industry developed in the course of the last 60 years from its very infancy to one of the biggest global markets. This tremendous evolution was triggered by several historical milestones. In 1947, John Bardeen and Walter Brattain presented the world's first transistor [1] based on Walter Shockley's field-effect theory. Their discovery was soon after rewarded by the Nobel Prize in physics and led to the development of today's semiconductor industry.

Another groundbreaking discovery was made in 1977, when Alan Heeger, Hideki Shirakawa, and Alan MacDiarmid presented the first conducting polymer [2]. Their work opened the way for organic semiconductors, which stand for cheap and flexible electronics like organic LEDs, photovoltaic cells, and field-effect transistors. With still a lot of ongoing fundamental research, some fields already reached maturity. Especially, organic LEDs became an irreplaceable part of modern televisions in the last couple of years. The major impact of organic semiconductors was awarded by Royal Swedish Academy of Sciences with the Nobel Price in chemistry.

On decade later, in 1988, Peter Grünberg and Albert Fert reported an effect, which they called the giant magneto resistance (GMR) [3, 4]. In contrary to conventional electronic devices, which use charges as carriers of information, the GMR exploits the electronic spin degree of freedom. Their discovery led to the development of a completely new branch of research, which is these days referred to as spintronics. With the success of data-storage industry, in the last 25 years, devices using the GMR effect became a part of our everyday live.

The drive for steady innovation led researchers to think about new devices which unify these great ideas and would, therefore, be even more performing. The famous article of Datta and Das in 1990 [5] was the first step towards a new age of spintronic devices. Their proposal described a transistor, which could amplify signals using spins currents only. However, for this transistor to work, efficient spin-polarization, injection, and long relaxation times are necessary. Especially, the relaxation time is usually limited by spin-orbit coupling and the hyperfine interaction.

© Springer International Publishing Switzerland 2016
S. Thiele, *Read-Out and Coherent Manipulation of an Isolated Nuclear Spin*,
Springer Theses, DOI 10.1007/978-3-319-24058-9_1

Fig. 1.1 Spin-relaxation
time τ_s versus spin-diffusion
length l_s. Organic
semiconductors are situated
in the upper left corner
corresponding to long
spin-lifetimes but short
diffusion lengths. The figure
was taken from [6], and the
used references correspond
to the ones from [6]

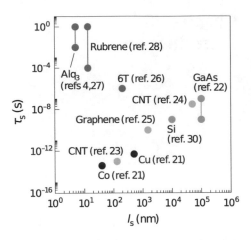

In this regard, organic spintronic devices might be a solution. They are known for
their intrinsically small spin scattering, which allows for long spin relaxation times
(see Fig. 1.1). This is because of the tiny spin-orbit interaction in organic materials.
The latter is proportional to Z^4, with Z being the atomic number, which makes spin
scattering very weak in carbon based devices.

In this context, single-molecule magnets (SMMs) are interesting candidates as building blocks for organic spintronic devices [7, 8]. Each molecule consists of a magnetic
core, which is surrounded by organic ligands. The latter do not only protect the core
from environmental influence but also tailor its magnetic properties. Replacing or
modifying the ligands by means of organic chemistry alters the environmental coupling and makes selective bonding to specific surfaces possible [9]. Likewise, one
can change the magnetic core, consisting of usually one or a few transition metal
or rare earth ions, to alter the spin system, the spin-orbit coupling, or the hyperfine
interaction of the molecule. Moreover, it is rather straight forward to synthesize bil-
lions of identical copies and embed them in virtually any matrix without changing
their magnetic properties. It is this versatility, which makes them very attractive for
spintronic devices.

The first, and most prominent, single-molecule magnet is the Mn_{12} acetate, which
was discovered by Lis in 1980 [15]. It consists of 12 manganese atoms, which are
surrounded by acetate ligands (see Fig. 1.2a). Another very famous single-molecule
magnet is the Fe_8 [16], consisting of eight iron(III) ions surrounded by a macrocyclic
ligand (see Fig. 1.2b). Both systems posses a total spin of $S = 10$ with an Ising
type anisotropy resulting in an energy barrier separating the $m_s = \pm 10$ ground
states by 63 K for Mn_{12} acetate [17] and by 25 K for Fe_8 [18]. In 1996, researches
found the first evidence of quantum properties in SMM crystals. It was observed
that the magnetization of the crystal is able to change its orientation via a tunnel
process [19, 20]. A few years later, it was discovered that quantum inference during
the tunnel process is possible [13]. And more recently, the coherent manipulation of

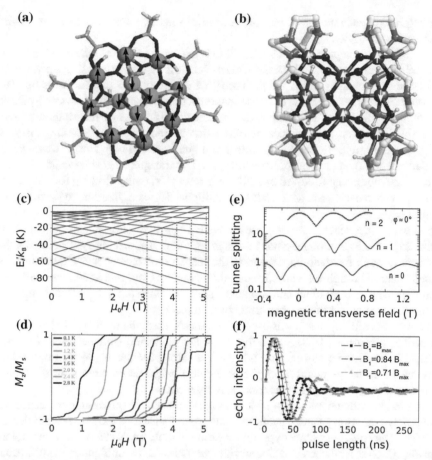

Fig. 1.2 a The Mn_{12} acetate SMM consists of 8 Mn(III) atoms with $S = 2$ (*orange*) and 4 Mn(IV) with $S = 3/2$ (*green*), which are connected via oxygen bonds. The spin of the twelve Mn atoms adds up to $S = 10$. Adapted from [10]. **b** The Fe_8 SMM, consists of eight Fe(III), which are interconnected by oxygen atoms (*red*). Each Fe(III) has spin of 5/2, which adds up to a total spin $S = 10$. Adapted from [11]. **c** Zeeman diagram of the Mn_{12} acetate obtained by exact numerical diagonalization. Important avoided level crossings are indicated by red dotted lines. **d** Magnetic hysteresis measurements obtained via Hall bar measurements of a microcrystal of $Mn_{12}tBuAc$. Adapted from [12]. **e** Quantum interference measurements obtained with a Fe_8 micro crystal. Adapted from [13]. **f** Rabi oscillations of a Fe_4 nano crystal. Adapted from [14]

the SMM's magnetic moment has been achieved for crystalline assemblies of SMMs [14, 21, 22].

The success of single-molecule magnets led to the discovery of a huge variety of new systems. A property which most of the experiments with SMMs have in common, is the use of a macroscopic amount of molecules in order to increase the detectable magnetic signal. However, a complete new type of experiments is possible when the molecules are measured isolated. Therefore, during the last couple of years, a lot of effort was put into the construction of ultra sensitive detectors towards single-molecule sensitivity.

A promising concept to study isolated SMMs makes use of spin-polarized scanning-tunneling spectroscopy [23]. Therein, the molecule is deposited on a single crystalline metallic surface and studied via the tunnel current through a tiny movable tip. The advantage of this technique is the combination of transport measurements with atomic resolution imaging, which makes an explicit identification of the studied system possible. However, the electrical manipulation by means of a gate voltage is hard to implement, and consequently, this technique comes along with a tremendous reduction of the amount of information gained by transport measurements.

Therefore, our group followed two different strategies, both, allowing for the implementation of a back-gate, which adds an additional degree of freedom to the transport measurements.

In the first approach, two molecules were deposited onto a carbon nanotube [24, 25]. Due to a strong exchange coupling, the first molecule spin polarizes the current through the nanotube, whereas the second molecule acts as a detector. The conductance through the carbon nanotube is larger if the molecules were aligned parallel, with respect to an antiparallel alignment. This spin valve effect leads to a magneto resistance change of several hundred percent.

The second method, which was used in this thesis, traps the molecule in between to metallic electrodes, thus, creating a single-molecule magnet spin-transitor [26, 27]. The tunnel current through the transistor becomes again spin dependent due to the exchange coupling of the molecule's magnetic moment with the tunnel current, giving rise to an all electrical spin read-out.

However, in both techniques, a lack of imaging makes the unambiguous identification of the SMM very hard. That is why our group focused on terbium double-decker SMMs. They possess a large hyperfine splitting of molecule's electronic ground state levels, which can be used as a fingerprint and makes an unambiguous identification even without imaging possible. Moreover, the strong hyperfine interaction allows for the read-out of a single nuclear spin [26]. The latter is well protected from the environment and, therefore, a promising candidate for quantum information processing.

1.2   Quantum Information Processing

The construction of a quantum computer is one of the most ambitious goals of today's scientists. The idea was already born in 1982, when Richard Feynman stated that certain quantum mechanical effects cannot be simulated efficiently with classical computers [28]. Three years later, David Deutsch was the first who demonstrated that quantum computers are outperforming classical computers regarding certain problems [29], but concrete algorithms to program such a computer remained scarce. The beginning of a widespread interest in quantum computation was triggered by Peter Shor in the mid 90's (see Fig. 1.3). He presented a quantum prime factorization algorithm, which exponentially outperformed any classical algorithm [30]. Two

Fig. 1.3 Number of citations
in Nature and Science whose
topic contained quantum
computing. Numbers were
taken from Web of Science

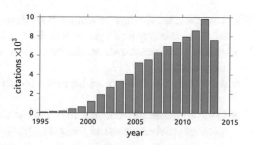

years later, Grover demonstrated that using a quantum computer to find an element
within an unsorted list would gain a polynomial speedup with respect to a classical
computer [31].

In analogy to the classical bit, the smallest processing unit of a quantum computer
is a quantum bit or qubit. It consists of a two level quantum system, whose states are
usually denoted as $|0\rangle$ and $|1\rangle$. The difference to a classical bit, which can be either
in 0 or 1, is that the qubit can be in the state $|0\rangle$, $|1\rangle$, or a superposition of both. This
superposition state is mathematically described as $a|0\rangle + b|1\rangle$. In order to visualize
a qubit, people often refer to the Bloch sphere (see Fig. 1.4). Therein, the $|0\rangle$ state
corresponds to the north pole and the $|1\rangle$ state to the south pole of the sphere. In
contrary to the classical bit, which is either at the north or the south pole, the qubit
state can be at any point of the sphere, corresponding to a superposition state.

The real power of a quantum computer is believed to be in its exponential growth of
the state space with increasing number of qubits. In contrary to a classical computer,
which is able to address $2n$ different states with n bits, a quantum computer can
address 2^n states with n bits.

Yet, to harness this power a real physical implementation of a quantum computer
is necessary. In order to decide whether or not a quantum mechanical system is
suited for constructing a quantum computer, DiVincenzo formulated the following
five criteria [32].

Fig. 1.4 Bloch sphere
representation of a quantum
bit. The two levels of a qubit
$|0\rangle$ and $|1\rangle$ are represented
by the north pole and the
south pole of the sphere and
any linear superposition can
be visualized as a point on
the surface of the sphere

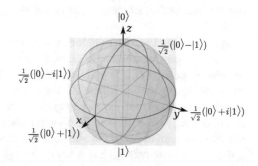

- **Information storage on qubits**: the information is encoded on a quantum property of a scalable physical system which lives long enough to perform computations.
- **Initial state preparation**: the state of the qubit needs to be prepared before each computation.
- **Isolation**: the qubit must be protected from decoherence by isolation from the environment.
- **Gate implementation**: the manipulation of a quantum state must be performed with reasonable precision and much faster than the decoherence time T_2.
- **Read-out**: the final state of the qubit must be read-out with a sufficiently high precision.

One of the most delicate criteria for any quantum mechanical system is the isolation from the environment.

One of the earliest experiments fulfilling these criteria was performed in the group of David Wineland [33]. To create a qubit they were using electrically trapped ions, which were isolated from the environment using a ultra-high vacuum (see Fig. 1.5a). In another approach the group of Serge Haroche trapped light inside a cavity with an extremely high quality factor (see Fig. 1.5b). Using the light matter interaction they could read-out the quantum state of a photon. Both Wineland and Haroche were awarded the Nobel Prize in physics in 2012.

Yet, both techniques are experimentally very demanding. In order to get an easier access to a qubit system, researches were looking for solid state qubit systems which can be made using standard nano-fabrication techniques. A very promising candidate are Josephson junctions coupled to superconducting resonators [39, 40]. However, their size of several μm makes them extremely sensitive to external noise.

Another possibility to create qubits follows the proposal of Loss and DiVincenzo [41] (see Fig. 1.5d). Therein, the spin of electron inside a quantum dot is used as a two level quantum system. Since they are much smaller than superconducting circuits, they couple less strongly to the environment, but at the same time they are also harder to detect. The first single-shot read-out of an electron spin inside a quantum dot was reported in 2004 [42]. One year later, Stotz et al. demonstrated the coherent transport of an electron spin inside a semiconductor [43], and in 2006, the coherent manipulation of an electron spin in a GaAs quantum dot was presented by Koppens et al. [44].

Despite their big success, the coupling to the environment is still sufficiently strong to destroy coherence within several hundred nanoseconds. Alternative concepts propose the use of nuclear spins as building blocks for quantum computing since they benefit from inherently longer coherence times compared to electronic spins, because of a better isolation from the environment. But weak coupling comes at a price: the detection and manipulation of individual nuclear spins remain challenging tasks.

Despite the difficulties, scientists demonstrated operating nuclear spin qubits using optical detection of nitrogen vacancy centers [45] (Fig. 1.5e), or by performing single-shot electrical measurements in silicon based devices [46] (Fig. 1.5f) and single-molecule magnet based devices [25, 26, 47] (Fig. 1.5g).

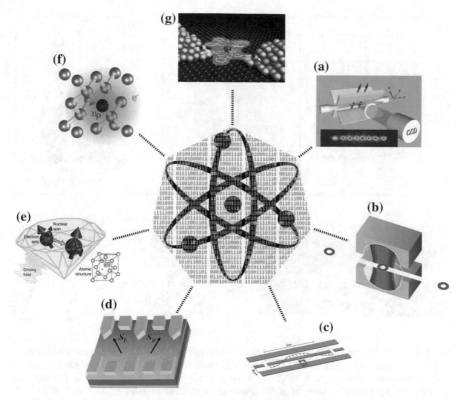

Fig. 1.5 Collection of different qubit types **a** ion traps taken from [34], **b** photons in a cavity source: Nobel Price Commite, **c** superconducting ciruits taken from [35], **d** quantum dots, source: Vitaly Golovach, **e** diamond color centers taken from [36], **f** ^{31}P impurities in silicon taken from [37] and **g** molecular magnets taken from [38]. Notice that selections focused on some important qubit families and is not a complete overview of all existing qubits

In order to solve the detection problem, the nuclear spin was measured indirectly through the hyperfine coupling to an electronic spin. Figure 1.6a explains this detection scheme exemplary using a NV defect, a color center in diamond [45]. The orbital ground state and the first excited state of the NV-center are $S = 1$ triplet states. Due to spin-spin interactions both states are split into a lower energy $m_s = 0$ ($|0_e\rangle$) state and two higher energy $m_s = \pm 1$ ($|\pm 1_e\rangle$) states. Their separation at zero magnetic field are 2.87 and 1.43 GHz for the ground state and excited state respectively. Optical transitions in NV-centers are spin preserving, leading to $\Delta m_s = 0$. If the spin is in the $m_s = 0$ ($m_s = \pm 1$) ground state, it can only be excited in the $m_s = 0$ ($m_s = \pm 1$) excited state. The average lifetime of the excited state is about 10 ns. After this time, a relaxation in the corresponding ground state takes place under the emission of a photon. If, however, the system was in the $m_s = \pm 1$ excited state, a relaxation via a non radiating metastable state into the $m_s = 0$ ground state is possible, causing a considerably smaller luminescence. The $|0_e\rangle$ and the $|\pm 1_e\rangle$ state are therefore

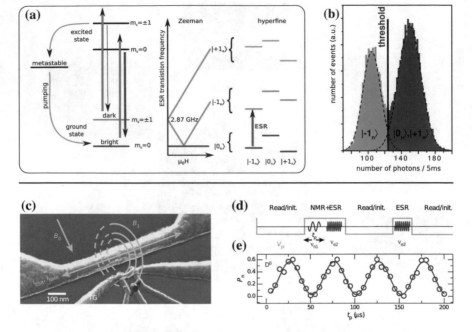

Fig. 1.6 **a** Energy diagrams of an NV-center. The left graph depicts radiative (*green and red arrows*) and non-radiative (*grey arrows*) transitions between the electronic ground state and the first excited state. In the center of the graph, the Zeeman diagram of the ground state triplet and its fine structure splitting were presented. The right graph shows the hyperfine splitting of each electronic state. Adapted from [45]. **b** Photon-counts histogram showing two Gausian like peaks. The left peak corresponds to the $|-1_n\rangle$ state and the right peak to the $|0_n\rangle$ and $|+1_n\rangle$ states. Adapted from [45]. **c** Scanning electron micrograph of a Si qubit. **d** Pulse signal of a coherent nuclear spin rotation and the subsequent read-out. **e** Rabi oscillations of a single ^{31}P nuclear spin. **c–e** were taken from [46]

referred to as the bright and the dark state, respectively. This enables the optical detection of the magnetic resonance (ODMR). Furthermore, the relaxation process via the metastable state is pumping the system into the $|0_e\rangle$ state, which is used to prepare the electronic spin in its initial state. The transition frequency between $|0_e\rangle \rightarrow |-1_e\rangle$ and $|0_e\rangle \rightarrow |+1_e\rangle$ can be changed by applying an external magnetic field along the quantization axis of the NV-center (see middle graph in Fig. 1.6a). Additionally, the hyperfine coupling to the nitrogen isotope ^{14}N, with a nuclear spin of $I = 1$, splits each electronic spin state into three, resulting in a nuclear spin dependent transition frequency under the influence of any external magnetic field. The three nuclear spin states will be referred to as $|-1_n\rangle$, $|0_n\rangle$, $|+1_n\rangle$. To detect the nuclear spin, the system is first pumped into the $|0_e\rangle$ state using a strong laser pulse. Afterward, a microwave pulse of precise duration and frequency is applied. If the frequency is matched to the $|0_e\rangle|-1_n\rangle \rightarrow |-1_e\rangle|-1_n\rangle$ level spacing, the electronic state will change from the bright into the dark state only if the nuclear spin was in the $|-1_n\rangle$ state (see left graph in Fig. 1.6a). The read-out is done by repeating this procedure several times and recording the luminescence signal. If the nuclear spin was in the $|0_n\rangle$ or $|+1_n\rangle$ state, the luminescence signal is larger than for the $|-1_n\rangle$ state (see Fig. 1.6b). Note that the detection of the nuclear spin state was realized by the read-out of the electronic spin state.

Quite similar to nitrogen color centers in diamond are ^{31}P impurities in silicon. However, owing to the small band gap of silicon the detection can be done electrically via a coupling to a close by quantum dot [46]. Notice that the nuclear spin read-out is again performed by exploiting the nuclear spin dependent electron spin resonance (ESR). Since the magnetic moment of the nuclear spin μ_N is about 2000 times smaller than the magnetic moment of the electronic spin μ_B, the manipulation of the former happens at times scales which are three orders of magnitude longer. In order to achieve a proper manipulation, large local AC magnetic fields are necessary. The group of Morello realized these fields by on-chip microwave strip lines (see Fig. 1.6c). The nuclear spin manipulation happened according to the following protocol (see Fig. 1.6d). First, the nuclear spin was prepared in its initial state. Afterward, a microwave pulse at the nuclear spin transition frequency of duration τ_p was applied. Depending on the pulse duration, the nuclear spin can be flipped with the probability P_n. Plotting P_n versus τ_p resulted in coherent Rabi oscillations (see Fig. 1.6e).

Nevertheless, the time scale of a manipulation remained in the order of $100~\mu s$ due to the tiny magnetic moment of the nuclear spin [46, 48]. Larger local alternating magnetic fields would increase this frequency, but they are difficult to generate using state of the art on-chip coils [49] due to the inevitable parasitic crosstalk to the detector and neighboring spin qubits.

To solve this problem, we propose and demonstrate in this thesis the single nuclear spin manipulation by means of an AC electric field. Indeed, it was already suggested by Kane [50] that the Stark effect of the hyperfine coupling could be used to tune different ^{31}P nuclear spins in and out of resonance using local DC gate voltages. He, therefore, established the individual addressability by applying only a global microwave field.

Our approach can be viewed as the extension of Kane's proposal to AC gate voltages. We will demonstrate coherent nuclear qubit manipulations using the hyperfine Stark effect to transform local electric fields into effective AC magnetic fields in the order of a few hundred mT and, hence, speeding up the clock speed of a single nuclear spin operation by two orders of magnitudes. In addition, we show that a local static gate voltage can shift the resonance frequency by several MHz, allowing for the individual addressability of several nuclear spin qubits.

1.3 Thesis Outline

My thesis was dedicated to study the read-out and manipulation of an isolated nuclear spin inside a single-molecule magnet. We made use of a three terminal transistor layout, in which the nuclear spin is electrically detected using a read-out quantum dot.

In order to give the reader a basic understanding of how the molecular spin-transistor works, we will recall in chapter two some fundamental transport properties of a quantum dot. In particular, we will focus on single electron tunneling, co-tunneling, and the Kondo effect since they are the most important transports characteristics observed in our devices.

In Chap. 3 we will concentrate on the magnetic properties of an isolated $TbPc_2$ single-molecule magnet. A lot of attention is directed to the electronic states of terbium ion, which are responsible for the observed magnetic properties of the device and, therefore, of paramount importance for this thesis.

A large part of my work was also devoted to the design and the construction of the experimental setup and is shown in chapter four. Starting from the dilution refrigerator I will explain each important part of the experiment which was added or modified in order to fabricate and measure a molecular spin-transistor.

Chapter 5 starts with explaining the mode of operation of the single-molecule magnet spin-transistor based on a simple model. The rest of the chapter details the conducted experiments in order to substantiate the aforementioned model.

In Chap. 6 we will use the spin-transistor to perform a time-resolved, quantum non-demolition read-out of the nuclear spin qubit state. We determined the relaxation time T_1 and the fidelity of the read-out. Furthermore, the experimental results are compared with quantum Monte Carlo simulations in order to deduce the dominating relaxation mechanism.

In Chap. 7 we propose and present the coherent manipulation of a single nuclear spin by means of the hyperfine Stark effect. Hence, using an AC electric field we generated and effective alternating magnetic field in the order of a few hundred mT. These results represent the first manipulation of a nuclear spin inside a single-molecule magnet and the first electrical manipulation of an isolated nuclear spin qubit.

References

1. J. Bardeen, W. Brattain, The transistor. A semi-conductor triode. Phys. Rev. **74**, 230–231 (1948)
2. C. Chiang, C. Fincher, Y. Park, A. Heeger, H. Shirakawa, E. Louis, S. Gau, A. MacDiarmid, Electrical conductivity in doped polyacetylene. Phys. Rev. Lett. **39**, 1098–1101 (1977)
3. J. Barnaś, A. Fuss, R. Camley, P. Grünberg, W. Zinn, Novel magnetoresistance effect in layered magnetic structures: theory and experiment. Phys. Rev. B **42**, 8110–8120 (1990)
4. M.N. Baibich, J.M. Broto, A. Fert, F.N. van Dau, F. Petroff, Giant magnetoresistance of (001)Fe/(001)Cr magnetic superlattices. Phys. Rev. Lett. **61**, 2472–2475 (1988)
5. S. Datta, B. Das, Electronic analog of the electro-optic modulator. Appl. Phys. Lett. **56**, 665 (1990)
6. G. Szulczewski, S. Sanvito, M. Cocy, A spin of their own. Nat. Mater. **8**, 693–5 (2009)
7. D. Gatteschi, R. Sessoli, J. Villain, *Molecular Nanomagnets* (Oxford University Press, Oxford, 2006)
8. L. Bogani, W. Wernsdorfer, Molecular spintronics using single-molecule magnets. Nat. Mater. **7**, 179–86 (2008)

9. S. Klyatskaya, J.R.G. Mascarós, L. Bogani, F. Hennrich, M. Kappes, W. Wernsdorfer, M. Ruben, Anchoring of rare-earth-based single-molecule magnets on single-walled carbon nanotubes. J. Am. Chem. Soc. **131**, 15143–51 (2009)
10. R. Sessoli, H.L. Tsai, A.R. Schake, S. Wang, J.B. Vincent, K. Folting, D. Gatteschi, G. Christou, D.N. Hendrickson, High-spin molecules: [Mn12O12(O2CR)16(H2O)4]. J. Am. Chem. Soc. **115**, 1804–1816 (1993)
11. W. Wernsdorfer, Molecular nanomagnets: towards molecular spintronics. Int. J. Nanotechnol. **7**, 497 (2010)
12. W. Wernsdorfer, M. Murugesu, G. Christou, Resonant tunneling in truly axial symmetry Mn12 single-molecule magnets: sharp crossover between thermally assisted and pure quantum tunneling. Phy. Rev. Lett. **96**, 057208 (2006)
13. W. Wernsdorfer, Quantum phase interference and parity effects in magnetic molecular clusters. Science **284**, 133–135 (1999)
14. C. Schlegel, J. van Slageren, M. Manoli, E.K. Brechin, M. Dressel, Direct observation of quantum coherence in single-molecule magnets. Phys. Rev. Lett. **101**, 147203 (2008)
15. T. Lis, Preparation, structure, and magnetic properties of a dodecanuclear mixed-valence manganese carboxylate. Acta Crystallogr. B Struct. Crystallogr. Cryst. Chem. **36**, 2042–2046 (1980)
16. K. Weighardt, K. Pohl, I. Jibril, G. Huttner, Hydrolysis Products of the Monomeric Amine Complex $(C_6H_{15}N_3)FeCl_3$: The Structure of the Octameric Iron(III) Cation of $\{[(C_6H_{15}N_3)_6Fe_8(\mu_3\text{-}O)_2(\mu_2\text{-}OH)_{12}]Br_7(H_2O)\}Br \cdot 8H_2O$. Angew. Chem. Int. Ed. Engl. **23**, 77–78 (1984)
17. A. Caneschi, D. Gatteschi, R. Sessoli, A.L. Barra, L.C. Brunel, M. Guillot, Alternating current susceptibility, high field magnetization, and millimeter band EPR evidence for a ground S = 10 state in [Mn12O12(Ch3COO)16(H2O)4].2CH3COOH.4H2O. J. Am. Chem. Soc. **113**, 5873–5874 (1991)
18. A.-L. Barra, P. Debrunner, D. Gatteschi, C.E. Schulz, R. Sessoli, Superparamagnetic-like behavior in an octanuclear iron cluster. Europhys. Lett. (EPL) **35**, 133–138 (1996)
19. J.R. Friedman, M.P. Sarachik, R. Ziolo, Macroscopic measurement of resonant magnetization tunneling in high-spin molecules. Phys. Rev. Lett. **76**, 3830–3833 (1996)
20. L. Thomas, F. Lionti, R. Ballou, D. Gatteschi, R. Sessoli, B. Barbara, Macroscopic quantum tunnelling of magnetization in a single crystal of nanomagnets. Nature **383**, 145–147 (1996)
21. S. Bertaina, S. Gambarelli, A. Tkachuk, I.N. Kurkin, B. Malkin, A. Stepanov, B. Barbara, Rare-earth solid-state qubits. Nat. Nanotechnol. **2**, 39–42 (2007)
22. A. Ardavan, O. Rival, J. Morton, S. Blundell, A. Tyryshkin, G. Timco, R. Winpenny, Will spin-relaxation times in molecular magnets permit quantum information processing? Phys. Rev. Lett. **98**, 1–4 (2007)
23. J. Schwöbel, Y. Fu, J. Brede, A. Dilullo, G. Hoffmann, S. Klyatskaya, M. Ruben, R. Wiesendanger, Real-space observation of spin-split molecular orbitals of adsorbed single-molecule magnets. Nat. Commun. **3**, 953 (2012)
24. M. Urdampilleta, S. Klyatskaya, J.-P. Cleuziou, M. Ruben, W. Wernsdorfer, Supramolecular spin valves. Nat. Mater. **10**, 502–6 (2011)
25. M. Ganzhorn, S. Klyatskaya, M. Ruben, W. Wernsdorfer, Strong spin-phonon coupling between a single-molecule magnet and a carbon nanotube nanoelectromechanical system. Nat. Nanotechnol. **8**, 165–169 (2013)
26. R. Vincent, S. Klyatskaya, M. Ruben, W. Wernsdorfer, F. Balestro, Electronic read-out of a single nuclear spin using a molecular spin transistor. Nature **488**, 357–360 (2012)
27. E. Burzurí, A.S. Zyazin, A. Cornia, H.S.J. van der Zant, Direct observation of magnetic anisotropy in an individual Fe_4 single-molecule magnet. Phys. Rev. Lett. **109**, 147203 (2012)
28. R.P. Feynman, Simulating physics with computers. Int. J. Theoret. Phys. **21**, 467–488 (1982)
29. D. Deutsch, Quantum theory, the church-turing principle and the universal quantum computer. Proc. R. Soc. A Math. Phys. Eng. Sci. **400**, 97–117 (1985)
30. P. Shor, Algorithms for quantum computation: discrete logarithms and factoring, in *Proceedings 35th Annual Symposium on Foundations of Computer Science*, vol. 26 (IEEE Computer Society Press, 1994), pp. 124–134. ISBN 0-8186-6580-7

31. L.K. Grover, A fast quantum mechanical algorithm for database search, in *Proceedings of the Twenty-Eighth Annual ACM Symposium on Theory of Computing—STOC '96* (ACM Press, New York, New York, USA, 1996). ISBN 0897917855

32. D.P. DiVincenzo, *Topics in Quantum Computers*, vol. 345. NATO Advanced Study Institute. Ser. E Appl. Sci. vol. 345 (1996)

33. D. Wineland, R. Drullinger, F. Walls, Radiation-pressure cooling of bound resonant absorbers. Phys. Rev. Lett. **40**, 1639–1642 (1978)

34. R. Blatt, D. Wineland, Entangled states of trapped atomic ions. Nature **453**, 1008–15 (2008)

35. J. Clarke, F.K. Wilhelm, Superconducting quantum bits. Nature **453**, 1031–42 (2008)

36. S.C. Benjamin, J.M. Smith, Driving a hard bargain with diamond qubits. Physics **4**, 78 (2011)

37. A. Morello, Quantum information: atoms and circuits unite in silicon. Nat. Nanotechnol. **8**, 233–4 (2013)

38. S. Thiele, R. Vincent, M. Holzmann, S. Klyatskaya, M. Ruben, F. Balestro, W. Wernsdorfer, Electrical readout of individual nuclear spin trajectories in a single-molecule magnet spin transistor. Phys. Rev. Lett. **111**, 037203 (2013)

39. L. Dicarlo, M.D. Reed, L. Sun, B.R. Johnson, J.M. Chow, J.M. Gambetta, L. Frunzio, S.M. Girvin, M.H. Devoret, R.J. Schoelkopf, Preparation and measurement of three-qubit entanglement in a superconducting circuit. Nature **467**, 574–8 (2010)

40. M. Neeley, R.C. Bialczak, M. Lenander, E. Lucero, M. Mariantoni, A.D. O'Connell, D. Sank, H. Wang, M. Weides, J. Wenner, Y. Yin, T. Yamamoto, A.N. Cleland, J.M. Martinis, Generation of three-qubit entangled states using superconducting phase qubits. Nature **467**, 570–3 (2010)

41. D. Loss, D.P. DiVincenzo, Quantum computation with quantum dots. Phys. Rev. A **57**, 120–126 (1998)

42. J.M. Elzerman, R. Hanson, L.H. Willems Van Beveren, B. Witkamp, L.M.K. Vandersypen, L.P. Kouwenhoven, Single-shot read-out of an individual electron spin in a quantum dot. Nature **430**, 431–5 (2004)

43. J.A.H. Stotz, R. Hey, P.V. Santos, K.H. Ploog, Coherent spin transport through dynamic quantum dots. Nat. Mater. **4**, 585–8 (2005)

44. F.H.L. Koppens, C. Buizert, K.J. Tielrooij, I.T. Vink, K.C. Nowack, T. Meunier, L.P. Kouwenhoven, L.M.K. Vandersypen, Driven coherent oscillations of a single electron spin in a quantum dot. Nature **442**, 766–71 (2006)

45. P. Neumann, J. Beck, M. Steiner, F. Rempp, H. Fedder, P.R. Hemmer, J. Wrachtrup, F. Jelezko, Single-shot readout of a single nuclear spin. Science (New York, N.Y.) **329**, 542–4 (2010)

46. J.J. Pla, K.Y. Tan, J.P. Dehollain, W.H. Lim, J.J.L. Morton, F.A. Zwanenburg, D.N. Jamieson, A.S. Dzurak, A. Morello, High-fidelity readout and control of a nuclear spin qubit in silicon. Nature **496**, 334–338 (2013)

47. M. Urdampilleta, S. Klyatskaya, M. Ruben, W. Wernsdorfer, Landau-Zener tunneling of a single Tb^{3+} magnetic moment allowing the electronic read-out of a nuclear spin. Phys. Rev. B **87**, 195412 (2013)

48. W. Pfaff, T.H. Taminiau, L. Robledo, H. Bernien, M. Markham, D.J. Twitchen, R. Hanson, Demonstration of entanglement-by-measurement of solid-state qubits. Nat. Phys. **9**, 29–33 (2012)

49. T. Obata, M. Pioro-Ladrière, T. Kubo, K. Yoshida, Y. Tokura, S. Tarucha, Microwave band on-chip coil technique for single electron spin resonance in a quantum dot. Rev. Sci. Instrum. **78**, 104704 (2007)

50. B.E. Kane, A silicon-based nuclear spin quantum computer. Nature **393**, 133–137 (1998)

Chapter 2
Single Electron Transistor

The first single electron transistor (SET), made of small tunnel junctions, was realized in the Bell Laboratories in 1987 by Fulton and Dolan [1]. Since then, the fabrication of SETs became more and more sophisticated and allowed for operation at room temperature [2] or as sensors for electron spin detection [3]. In this thesis, a single electron transistor will be used to read-out the state of an isolated nuclear spin and, therefore, a basic knowledge of the transport properties in SETs and its associated effects such as Coulomb blockade, elastic and inelastic cotunneling, and the Kondo effect are necessary.

2.1 Equivalent Circuit

A single electron transistor consists of a conducting island or quantum dot, which is tunnel-coupled to the source and drain leads. Due to the small size of the dot the electronic energy levels E_n are discretized. In order to observe the characteristic single electron tunneling through the device, the resistance R_t of the tunnel barriers should be much higher than the quantum of resistance:

$$R_t \gg \frac{h}{e^2} \tag{2.1.1}$$

where h is the Planck constant and e the elementary charge. This condition ensures that only one electron at the time is tunneling in or out of the quantum dot. A simple model to describe the electron transport through the dot was developed by Korotkov et al. [4], and reviewed by Kouwenhoven [5], and Hanson [6]. Therein, the quantum dot is coupled via constant source, drain, and gate capacitors (C_s, C_d, C_g) to the three

© Springer International Publishing Switzerland 2016
S. Thiele, *Read-Out and Coherent Manipulation of an Isolated Nuclear Spin*,
Springer Theses, DOI 10.1007/978-3-319-24058-9_2

Fig. 2.1 Equivalent circuit
of an SET. The electrostatic
behavior of the dot is
modeled by capacitors to the
source, drain, and gate
terminals

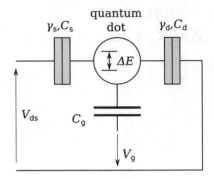

terminals as shown in Fig. 2.1. By applying a voltage to the three different terminals,
the electrostatic potential U_{es} of the quantum dot is modified as:

$$U_{es} = \frac{(C_s V_s + C_d V_d + C_g V_g)^2}{2 C_\Sigma} \tag{2.1.2}$$

with $C_\Sigma = C_s + C_d + C_g$ and V_s, V_d, and V_g being the source, drain, and gate voltages,
respectively. Furthermore, due to the Coulomb repulsion, adding an electron to the
quantum dot with N electrons ($N > 0$) will cost an additional energy:

$$U_c = \frac{E_c}{2} = \frac{e^2}{2 C_\Sigma} \tag{2.1.3}$$

with E_c being the charging energy. Accordingly, to observe single electron tunneling,
temperatures smaller than E_c are required since, otherwise, the tunnel process can
be activate thermally.

$$E_c \gg k_B T \tag{2.1.4}$$

Putting all contributions together results in the total energy U of the quantum dot
with N electrons:

$$U(N) = \frac{(-e(N - N_0) + C_s V_s + C_d V_d + C_g V_g)^2}{2 C_\Sigma} + \sum_1^N E_n(B) \tag{2.1.5}$$

where N_0 is the offset charge and $E_n(B)$ the magnetic field dependent single electron
energies. Experimentally, it is more convenient to work with the chemical potential,
defined as the energy difference between two subsequent charge states $\mu_{dot}(N) = U(N) - U(N - 1)$. Inserting Eq. 2.1.5 into this expression gives:

$$\mu_{dot}(N) = \left(N - \frac{1}{2}\right) E_c - \frac{E_c}{|e|}(C_s V_s + C_d V_d + C_g V_g) + E_N(B) \tag{2.1.6}$$

with E_N being the energy of the Nth electron in the quantum dot. Notice that the chemical potential depends linearly on the gate voltage, whereas the total energy shows a quadratic dependence. Therefore, the energy difference between the chemical potentials of different charge states remains constant for any applied voltages. The energy to add an electron to the quantum dot is called addition energy E_{add} and is defined as the difference between to subsequent chemical potentials.

$$E_{add}(N) = \mu(N+1) - \mu(N) = E_c + \Delta E \tag{2.1.7}$$

with ΔE being the energy spacing between two discrete energy levels.

2.2 Coulomb Blockade

The transport through the quantum dot is very sensitive to the alignment of the chemical potential μ inside the dot with respect to those of the source μ_s and drain μ_d. If we neglect the level broadening and any excited states of the quantum dot for a moment, then the transport through the SET can be explained with Fig. 2.2. Notice that V_{ds} and V_g are in arbitrary units and $V_g = 0$ when $\mu_{dot} = \mu_s = \mu_d$.

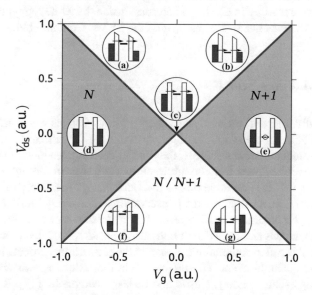

Fig. 2.2 Schematic of a stability diagram. Inside the *grey* regions the charge of the quantum dot is fixed to N (d) or $N + 1$ (e), leading to the Coulomb blockade. Likewise, inside the white area electrons can tunnel in and out of the quantum dot. If the conductance dI/dV is measured instead of the current I, only the *red* and the *blue line* are visible, corresponding to a change in I. Along the *red line* the chemical potential of the quantum dot is aligned with the source chemical potential, whereas along the *blue line* it is aligned with the drain chemical potential

First we want to discuss what happens for zero bias $V_{ds} = V_d = V_s = 0$. If $V_g < 0$, the chemical potential of the dot is larger than the chemical potential of the leads, and the SET is in its off state (Fig. 2.2d). Increasing V_g to zero will align the three chemical potentials. Electrons can tunnel in and out of the dot from both sides leading to a finite conductance and a charge fluctuation between N and $N + 1$. This particular working regime is called the charge degeneracy point (Fig. 2.2d). A further increase of V_g will push the chemical potential of the dot below the ones of source and drain, and the SET is again in its off state, but having $N + 1$ electrons on the dot. Whenever the charge of the dot is fixed, the SET is in the Coulomb blockade regime since adding another electron would cost energy to overcome the electron-electron repulsion.

If we now increase the bias voltage to $V_{ds} \neq 0$, we shift the chemical potential between source and drain and open an energy or bias window of $\mu_s - \mu_d = eV_{ds}$, and a current is observed even for $V_g \neq 0$.

The red line in Fig. 2.2 corresponds to the situation where the chemical potential of the dot is aligned with μ_s (Fig. 2.2a, g). Crossing this line will turn the SET on or off, resulting in a conductance ridge along the line. The slope can be calculated from the equivalent circuit by setting the potential difference between dot and source to zero and is given by $-C_g/(C_g + C_s)$.

On the other hand, if μ_{dot} is aligned with the drain chemical potential, the SET turns also on or off, resulting in another conductance ridge (blue line in Fig. 2.2). Its slope is of opposite sign and calculated by setting the potential difference between drain and dot to zero, resulting in C_g/C_d. Therefore, inside the white region the transistor is turned on, whereas inside the grey region the SET is Coulomb blocked.

2.3   Cotunneling Effect

Up to now only transport through energetically allowed states was considered. This is usually sufficient if the tunnel barrier resistances are larger than 1 MΩ. However, for smaller tunnel barrier resistances, the time to exchange an electron between the dot and the leads becomes fast enough to allow for transport through energetically forbidden states. This is possible due to the Heisenberg uncertainty relation, which states that a system can violate energy conservation within a very short time $\tau = \hbar/E$, where $E \approx E_c$ for quantum dots. Therefore within the time τ an electron can enter the quantum dot whereas another is tunneling into the leads. Since this process involves two electrons it is called cotunneling. Note that the entire tunnel process is considered to be a single quantum event. We distinguish in the following two different cases of cotunneling events, namely, elastic and inelastic cotunneling [7]. If the electron entering the quantum dot occupies the same energy level as the outgoing one, the cotunneling is elastic and requires no additional energy (Fig. 2.3a). Experimentally, it can be observed as a conductance background inside the Coulomb blocked region. If, however, the electron entering the dot occupies an excited state, separated by ΔE from the chemical potential of the electron leaving the dot, the transport is inelastic (Fig. 2.3b). This process requires energy and happens only at finite bias voltages with

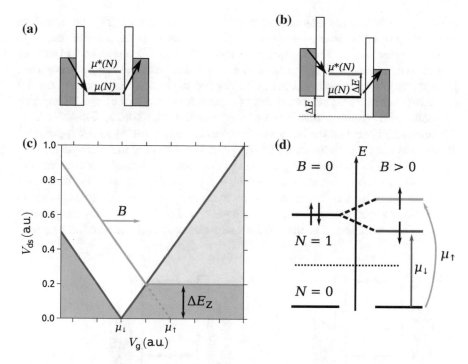

Fig. 2.3 Schematic showing the elastic **a** and inelastic **b** cotunneling process. **c** Stability diagram for a quantum dot whose ground and excited state are split by the Zeeman energy. The conductance step inside the Coulomb blocked region (*dark grey*) occurs at $e|V_{ds}| = \Delta E_Z$ and can be used to determine the Zeeman splitting. **d** Energy level diagram of a quantum dot with zero or one electron. Due to an external magnetic field, the degeneracy between *spin up* and *down* is lifted. The two lowest lying chemical potentials of the quantum dot correspond to the energy difference of the Zeeman split $N = 1$ doublet and the $N = 0$ singlet

$e|V_{ds}| > \Delta E$. The result is a conductance step inside the Coulomb blocked region. In the case of a very simple quantum dot as shown in Fig. 2.3c, d the conductance step can be used to determine the Zeeman splitting due to a magnetic field.

2.4 Kondo Effect

In the 1930s, de Haas et al. found out that while cooling down a long wire of gold, the resistance reaches a minimum at around 10 K and increases for further cooling [8]. Later, it was discovered that this effect was correlated to the presence of magnetic impurities, but a theoretical explanation of this phenomenon was only presented in the 1960s, by Jun Kondo [9]. In his model, an antiferromagnetic coupling between the conduction electrons and the residual magnetic impurities leads to the formation of a singlet state below a certain temperature T_K (Kondo temperature). This can be thought of a cloud of conduction electrons, screening the magnetic impurity and therefore augmenting its effective cross section, which causes an increase in resistance.

The same effect can be found in quantum dots. If they are filled with an odd number of electrons, its total spin $S = 1/2$, which makes it an artificial magnetic impurity. If, furthermore, the coupling between the dot and the leads is large enough (tunnel resistances below 1 MΩ), electrons from the leads try to screen the artificial impurity by continuously flipping its spin via a tunnel process (Fig. 2.4a, b). This allows for a hybridization between the leads and the quantum dot, resulting in the appearance of two peaks in the quantum dot's DOS: one at Fermi level of the source and one at the Fermi level of the drain (Fig. 2.4c). The conductance through the quantum dot can be explained by the convolution of the two peaks. Since at zero V_{ds} the source and drain Fermi level coincide, the conductance will have a maximum and drops to zero for higher bias voltages, resulting in a peak, or Kondo ridge. If the temperature becomes comparable to T_K, the antiferromagnetic coupling between the magnetic

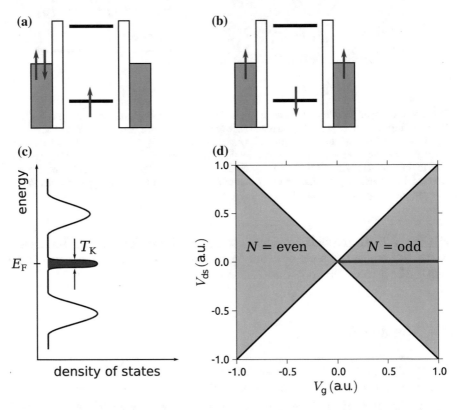

Fig. 2.4 Kondo transport mechanism with the initial state (**a**) and the final state (**b**). Note that the spin of the quantum dot flipped during the process. **c** The Kondo effect creates a peak in the density of states, whose width is given by the Kondo temperature. **d** Experimentally, the Kondo effect is observed as a conductance ridge or Kondo ridge (*red line*) inside the Coulomb blockade region of the stability diagram

impurity and the electrons in the leads is destroyed, resulting in the suppression of the conductance peak. The temperature dependance can be fitted by the empirical Goldhaber-Gordon equation [10]

$$G(T) = G_0 \left(\frac{T^2}{T_K^2} (2^{1/s} + 1) \right)^{-s} + G_c \qquad (2.4.1)$$

and results in $G(T_K) = G_0/2$. The variable G_c accounts for a conductance offset caused by elastic cotunneling and $s = 0.22$.

Another possibility to study the Kondo effect is by applying a magnetic field. For a classical spin 1/2 in a magnetic field, the Zeeman effect will split the spin up and down levels by $g\mu_B B_c$. If this splitting becomes larger than the antiferromagnetic coupling given by $0.5k_B T_K$ [11], the Kondo ridge at zero bias is destroyed. However, applying a positive or negative bias voltage V_{ds} can compensate for the energy gap when $e|V_{ds}| = g\mu_B|B|$. This leads to the revival of the Kondo effect and is observed as two peaks, one at negative and one at positive bias. The separation of the Kondo peak as a function of the applied magnetic field is schematically displayed in Fig. 2.5a.

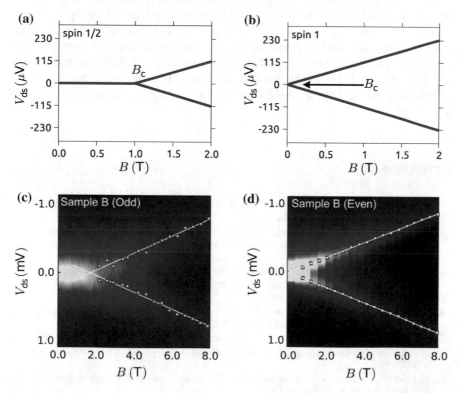

Fig. 2.5 Separation of the spin 1/2 (**a**) and spin 1 (**b**) Kondo peak as a function of the applied magnetic field. The critical field of 0.5 T was chosen arbitrarily. The slope corresponds corresponds to a g factor of 2. (**c, d**) Experimental data adapted from [12]

Using this model, the critical field can be used to estimate the Kondo temperature and the strength of the coupling.

If we add one electron to the quantum dot, the spin is either zero (singlet state) or one (triplet state). In case of zero spin, no Kondo effect will be observed. If, however, the triplet state becomes the ground state of the system, the situation changes. Similar to the spin 1/2 Kondo effect, electrons from the leads try to screen the artificial magnetic impurity, which now has a spin of 1. Therefore, the screening requires two conduction channels, one for each electron of the triplet.

In quantum dots, like the ones we used in our experiments, the coupling of different energy levels to the source and drain terminals is not symmetric in energy, resulting in two individual Kondo temperatures T_{K_1} and T_{K_2}. Hence, in the temperature window $T_{K_1} < T < T_{K_2}$ the screening of channel 1 is suppressed, whereas the screening of channel 2 is still working. This scenario is referred to as the underscreened Kondo effect. Its signature is such that the critical field needed to quench the conductance ridge is much smaller than $0.5k_B T_K$ [12]. This can be understood by a semi-classical consideration of the residual spin which was left unscreened. The ferromagnetic coupling between the two spins, which led to the formation of the triplet ground state, results in an effective magnetic field created by the unscreened spin at the site of the screen spin. This field weakens the antiferromagnetic coupling of the second spin to the electrons in the leads. Hence, already at very small external magnetic fields, the critical field is reached, leading to a shift of B_c towards zero (Fig. 2.5b). The magnitude of the shift is proportional to the ferromagnetic exchange coupling between the two spins but cannot be determined precisely due to the lack of knowledge of the fully screened B_c. However, it gives an estimate of its order of magnitude.

References

1. T. Fulton, G. Dolan, Observation of single-electron charging effects in small tunnel junctions. Phys. Rev. Lett. **59**, 109–112 (1987)
2. K. Matsumoto, M. Ishii, K. Segawa, Y. Oka, B.J. Vartanian, J.S. Harris, Room temperature operation of a single electron transistor made by the scanning tunneling microscope nanooxidation process for the TiOx/Ti system. Appl. Phys. Lett. **68**, 34 (1996)
3. J.M. Elzerman, R. Hanson, L.H. Willems Van Beveren, B. Witkamp, L.M.K. Vandersypen, L.P. Kouwenhoven, Single-shot read-out of an individual electron spin in a quantum dot. Nature **430**, 431–435 (2004)
4. A. Korotkov, D. Averin, K. Likharev, Single-electron charging of the quantum wells and dots. Phys. B Condens. Matter **165–166**, 927–928 (1990)
5. L.P. Kouwenhoven, C.M. Marcus, P.L. Mceuen, S. Tarucha, M. Robert, *Electron Transport in Quantum Dots* (Proceedings of the NATO Advanced Study Institute, Kluwer, 1997). ISBN 140207459X
6. R. Hanson, J.R. Petta, S. Tarucha, L.M.K. Vandersypen, Spins in few-electron quantum dots. Rev. Mod. Phys. **79**, 1217–1265 (2007)
7. M. Pustilnik, L. Glazman, Kondo effect in quantum dots. J. Phys. Condens. Matter **16**, R513–R537 (2004)

8. W. de Haas, J. de Boer, G. van dën Berg, The electrical resistance of gold, copper and lead at low temperatures. Physica **1**, 1115–1124 (1934)
9. J. Kondo, Resistance minimum in dilute magnetic alloys. Prog. Theor. Phys. **32**, 37–49 (1964)
10. D. Goldhaber-Gordon, J. Göres, M. Kastner, H. Shtrikman, D. Mahalu, U. Meirav, From the Kondo regime to the mixed-valence regime in a single-electron transistor. Phys. Rev. Lett. **81**, 5225–5228 (1998)
11. T.A. Costi, Kondo effect in a magnetic field and the magnetoresistivity of Kondo alloys. Phys. Rev. Lett. **85**, 1504–1507 (2000)
12. N. Roch, S. Florens, T.A. Costi, W. Wernsdorfer, F. Balestro, Observation of the underscreened Kondo effect in a molecular transistor. Phys. Rev. Lett. **103**, 197202 (2009)

Chapter 3
Magnetic Properties of TbPc$_2$

3.1 Structure of TbPc$_2$

The single molecule magnet which was investigated in this thesis is a metal-organic complex called bisphthalocyaninato terbium(III) ([TbPc$_2$]$^-$). The magnetic moment of the molecule arises from a single terbium ion (Tb^{3+}), situated in the center of the molecule. It is eightfold coordinate to the nitrogen atoms of the two phthalocyanine (Pc) ligands, which are stacked below and above the terbium ion resulting an approximate C_4 symmetry in the close environment of the Tb. The ligands are encapsulating the Tb^{3+} in order to preserve and tailor its magnetic properties. Its resemblance to the double-decker airplane of the 1920s is giving it its colloquial name—terbium double-decker (Fig. 3.1).

3.2 Electronic Configuration of Tb^{3+}

Naturally attained ^{159}Tb is one of the 22 elements with only one natural abundant isotope. With an atomic number of 65, it is situated within the lanthanide series in the periodic system of elements (see Fig. 3.2). Its name arises from the Swedish town Ytterby, where it was first discovered in 1843.

The electronic structure of Tb is [Xe]4f^{9}6s^2. The 4f shell, which is not completely filled, is responsible for its paramagnetism. It is located inside the 6s, 5s, and 5p shell and therefore well protected from the environment. Like most of the lanthanides, Tb releases three electrons to form chemical bonds. These three electrons consist of two 6s electrons, which are on the outer most shell and therefore easy to remove, and one 4f electron. 4f electrons are most of the time inside the 5s and 5p shell, but they cannot come very close to the core neither, resulting in a smaller ionization energy than for 5s and 5p electrons. Thus, the electronic structure of the Tb^{3+} is [Xe]4f^8.

The energetic position of the different orbits and levels of the terbium ion is affected by several interactions, namely, the electron-electron interaction H_{ee}, the spin-

© Springer International Publishing Switzerland 2016
S. Thiele, *Read-Out and Coherent Manipulation of an Isolated Nuclear Spin*,
Springer Theses, DOI 10.1007/978-3-319-24058-9_3

(a)   **(b)**

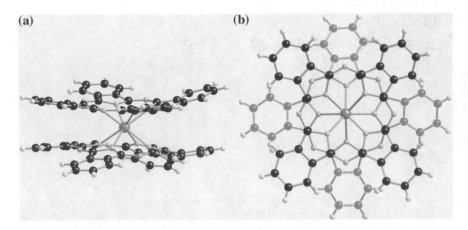

Fig. 3.1 *Side view* (**a**) and *top view* (**b**) of the TbPc$_2$. The *pink atom* in the center of the complex is the Tb^{3+} ion, which is eightfold coordinated to the nitrogen atoms (*blue*) of the two phthalocyanine ligands resulting in a local approximate C_4 symmetry.

1 **H** Hydrogen 1.00794																		2 **He** Helium 4.003
3 **Li** Lithium 6.941	4 **Be** Beryllium 9.012182											5 **B** Boron 10.811	6 **C** Carbon 12.0107	7 **N** Nitrogen 14.00674	8 **O** Oxygen 15.9994	9 **F** Fluorine 18.9984032	10 **Ne** Neon 20.1797
11 **Na** Sodium 22.989770	12 **Mg** Magnesium 24.3050											13 **Al** Aluminum 26.981538	14 **Si** Silicon 28.0855	15 **P** Phosphorus 30.973761	16 **S** Sulfur 32.066	17 **Cl** Chlorine 35.4527	18 **Ar** Argon 39.948
19 **K** Potassium 39.0983	20 **Ca** Calcium 40.078	21 **Sc** Scandium 44.955910	22 **Ti** Titanium 47.867	23 **V** Vanadium 50.9415	24 **Cr** Chromium 51.9961	25 **Mn** Manganese 54.938049	26 **Fe** Iron 55.845	27 **Co** Cobalt 58.933200	28 **Ni** Nickel 58.6934	29 **Cu** Copper 63.546	30 **Zn** Zinc 65.39	31 **Ga** Gallium 69.723	32 **Ge** Germanium 72.61	33 **As** Arsenic 74.92160	34 **Se** Selenium 78.96	35 **Br** Bromine 79.904	36 **Kr** Krypton 83.80
37 **Rb** Rubidium 85.4678	38 **Sr** Strontium 87.62	39 **Y** Yttrium 88.90585	40 **Zr** Zirconium 91.224	41 **Nb** Niobium 92.90638	42 **Mo** Molybdenum 95.94	43 **Tc** Technetium (98)	44 **Ru** Ruthenium 101.07	45 **Rh** Rhodium 102.90550	46 **Pd** Palladium 106.42	47 **Ag** Silver 107.8682	48 **Cd** Cadmium 112.411	49 **In** Indium 114.818	50 **Sn** Tin 118.710	51 **Sb** Antimony 121.760	52 **Te** Tellurium 127.60	53 **I** Iodine 126.90447	54 **Xe** Xenon 131.29
55 **Cs** Cesium 132.90545	56 **Ba** Barium 137.327	57 **La** Lanthanum 138.9055	72 **Hf** Hafnium 178.49	73 **Ta** Tantalum 180.9479	74 **W** Tungsten 183.84	75 **Re** Rhenium 186.207	76 **Os** Osmium 190.23	77 **Ir** Iridium 192.217	78 **Pt** Platinum 195.078	79 **Au** Gold 196.96655	80 **Hg** Mercury 200.59	81 **Tl** Thallium 204.3833	82 **Pb** Lead 207.2	83 **Bi** Bismuth 208.98038	84 **Po** Polonium (209)	85 **At** Astatine (210)	86 **Rn** Radon (222)
87 **Fr** Francium (223)	88 **Ra** Radium (226)	89 **Ac** Actinium (227)	104 **Rf** Rutherfordium (261)	105 **Db** Dubnium (262)	106 **Sg** Seaborgium (263)	107 **Bh** Bohrium (262)	108 **Hs** Hassium (265)	109 **Mt** Meitnerium (266)	110 (269)	111 (272)	112 (277)	113	114				

58 **Ce** Cerium 140.116	59 **Pr** Praseodymium 140.90765	60 **Nd** Neodymium 144.24	61 **Pm** Promethium (145)	62 **Sm** Samarium 150.36	63 **Eu** Europium 151.964	64 **Gd** Gadolinium 157.25	65 **Tb** Terbium 158.92534	66 **Dy** Dysprosium 162.50	67 **Ho** Holmium 164.93032	68 **Er** Erbium 167.26	69 **Tm** Thulium 168.93421	70 **Yb** Ytterbium 173.04	71 **Lu** Lutetium 174.967
90 **Th** Thorium 232.0381	91 **Pa** Protactinium 231.03588	92 **U** Uranium 238.0289	93 **Np** Neptunium (237)	94 **Pu** Plutonium (244)	95 **Am** Americium (243)	96 **Cm** Curium (247)	97 **Bk** Berkelium (247)	98 **Cf** Californium (251)	99 **Es** Einsteinium (252)	100 **Fm** Fermium (257)	101 **Md** Mendelevium (258)	102 **No** Nobelium (259)	103 **Lr** Lawrencium (262)

Fig. 3.2 Periodic table of elements. The element ^{159}Tb belongs to the lanthanide series and possesses only one stable isotope.

orbit coupling H_{so}, the ligand field potential H_{lf}, the exchange interaction H_{ex}, the hyperfine-coupling H_{hf}, and the magnetic field H_Z. An overview of the magnitude of these energetic effects on the 4f electrons is given in Table 3.1.

In the following we want to briefly discuss the different interactions starting with the Zeeman effect.

Table 3.1 Energy scale of different effects acting on 4f electrons.

Interaction	Energy equivalent (cm^{-1})
Electron-electron interaction H_{ee}	$\approx 10^4$
Spin-orbit coupling H_{so}	$\approx 10^3$
Ligand-field potential H_{lf}	$\approx 10^2$
Exchange interaction H_{ex}	< 1
Hyperfine interaction H_{hf}	$\approx 10^{-1}$
Magnetic field H_Z at 1T	≈ 0.5

Taken from [1, 2]

3.3 Zeeman Effect

From classical mechanics it is known that a magnetic moment μ exposed to an external magnetic field B will change its potential energy by $E_{pot} - -\mu B$. The quantum mechanical equivalent is called the Zeeman effect. To calculate the Zeeman energy we write down the Zeeman Hamiltonian:

$$H_Z = g\mu_B J B \tag{3.1}$$

where g is the Landée factor, $\mu_B = e\hbar/2m_e$ the Bohr magneton, and $J = L + S$ the total angular momentum of the system. For a more general derivation of this formula see Appendix A.1 and A.2. In the case of a free electron with $J = S$ and $B = (0, 0, B_z)$, the Zeeman Hamiltonian becomes:

$$H_Z = g\mu_B S_z B_z \tag{3.2}$$

with S_z being the Pauli matrix. Diagonalizing this Hamiltonian at different magnetic fields results in Fig. 3.3, which is referred to as the Zeeman diagram. It shows that the spin degeneracy if lifted at $B \neq 0$.

3.4 Electron-Electron Interaction

As we have seen in Table. 3.1, the electron-electron interaction is the strongest of all interactions and is mainly responsible for the orbital energies and the shell filling. The latter is well explained by the famous Hund's rules:

1. Hund's rule The electrons within a shell are arranged such that their total spin S is maximized. $\sum s_i \rightarrow max$. This can be understood in terms of Coulomb repulsion. Electrons with the same spin have to be in different orbitals due to the Pauli principle. Since they are in different orbitals, they are in average further apart from each other, resulting in a reduced Coulomb repulsion.

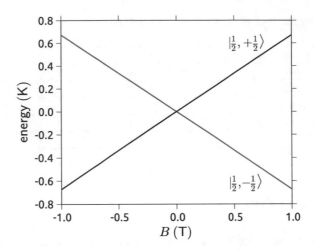

Fig. 3.3 Zeeman diagram of a free electron.

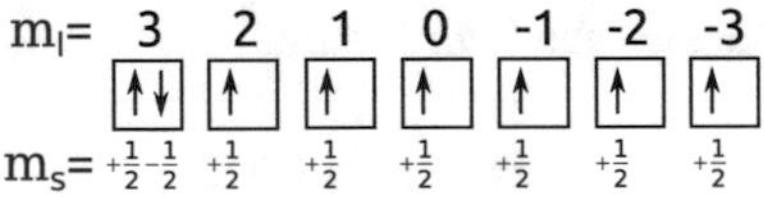

Fig. 3.4 Electronic structure of the Tb³⁺ 4f shell. $L = \sum_i m_l^i = 3$ and $S = \sum_i m_s^i = 3$

2. Hund's rule For a given spin, the electrons are arranged within the shell such that their total angular momentum L is maximized. $\sum l_i \rightarrow max$. This Hund's rule also origins from the Coulomb repulsion. Electrons with similar angular momentum are revolving more synchronous and avoiding each other therefore more effectively.

3. Hund's rule For less than half-filled sub-shells the total angular momentum $J = |L - S|$, whereas for more than half filled sub-shells the total angular momentum $J = |L + S|$. This rules arises from minimizing the spin-orbit coupling energy and cannot be explained easily with hand-waving arguments.

In order to fill up the 4f shell of Tb³⁺ we start with rule number one by putting seven electrons with spin up in the seven different orbitals and therefore maximize the spin S. The last electron is put in the $m_l = 3$ state according to the second rule. This already results in the final shell filling with a total spin $S = 7 \times \frac{1}{2} - \frac{1}{2} = 3$ and an angular momentum $L = 3 + 2 + 1 + 0 - 1 - 2 - 3 + 3 = 3$ as shown in Fig. 3.4.

3.5   Spin-Orbit Interaction

The spin-orbit interaction is the coupling of the electron's spin s with its orbital momentum l. In the semi-classical picture the electron's orbital motion creates a magnetic moment $\mu_l = -\frac{\mu_B}{\hbar} l$. Furthermore, since the famous Stern-Gerlach experiment from 1921, it is known that electrons possess a magnetic moment μ_s generated by its inherent spin s. Due to dipole interactions, the formerly independent momenta are connected, resulting in the total momentum $j = l + s$. The energy change resulting from this interaction is $\Delta E = -\mu_s B_l \propto ls$. Applying the correspondence principle leads to the spin-orbit Hamiltonian $H_{so} = \xi\, ls$, where ξ is the one-electron spin-orbit coupling parameter. A more exact derivation of the spin-orbit interaction is given in Appendix A.4 for the interested reader.

Since Tb^{3+} has eight electrons in the 4f shell we have to consider more than just one spin and orbital momentum. If, however, the coupling between different orbital momenta $H_{l_i-l_j} = a_{ij}l_i l_j$ and the different spins $H_{s_i-s_j} = b_{ij}s_i s_j$ is large compared to the spin-orbit coupling $H_{l_i s_i}$, the momenta itself couple first to a total spin $S = \sum_i s_i$ and a total orbital momentum $L = \sum_i l_i$, before coupling to the total momentum $J = L + S$, and the spin-orbit Hamiltonian modifies to:

$$H_{so} = \lambda(r)\, LS \tag{3.3}$$

Minimizing this energy for the Tb^{3+} leads to the third Hund's rule with a ground state of $J = L+S = 6$, which is $2J+1 = 13$ times degenerate. All possible combinations are displayed in Fig. 3.5. In the following paragraph we will see how the spin-orbit interaction can be computed within the framework of first order perturbation theory. Without spin-orbit coupling all spins would couple to the total spin S and all orbital momenta would combine to L, leading to $(2L+1) \times (2S+1)$ degenerate states. Since the spin-orbit contribution to the electron energy is small with respect to the electron-

Fig. 3.5 Due to the spin-orbit coupling, the total spin S, with its $2S + 1$ states, is coupling to the total orbital momentum L, with its $2L + 1$ states, resulting in a total momentum $J = L + S$ with $(2S + 1)(2L + 1)$ states.

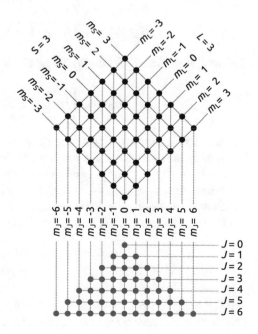

electron interaction, first-order degenerate perturbation theory can be applied. We perform our calculations using the L and S along with their projections m_L and m_S as good quantum numbers to describe our unperturbed states. To calculate the energy correction to first order for N degenerate states, we have to write the spin-orbit Hamiltonian in this basis as a $N \times N$ matrix and perform an exact numerical diagonalization. We assume that $\Psi_{u,v}^{(0)}$ is the unperturbed electron wave function and $u, v = [0..(2L + 1) \times (2S + 1)]$. Applying the product ansatz splits the wave function into a radial, angular and spin-dependent part: $\Psi_{u,v}^{(0)} = |R\rangle |m_L\rangle |m_S\rangle$. Since the operators L and S are not acting on the radial part we can write the Hamiltonian as:

$$H_{so} = \zeta \, LS \tag{3.4}$$

$$LS = L_x S_x + L_y S_y + L_z S_z \tag{3.5}$$

where $\zeta = \langle R | \lambda(r) | R' \rangle$ is the one-electron spin-orbit coupling constant. In order to expand this equation into a matrix, we make use of the following transformation:

$$L_x = \frac{1}{2}(L_+ + L_-); \quad L_y = \frac{1}{2i}(L_+ - L_-) \tag{3.6}$$

$$S_x = \frac{1}{2}(S_+ + S_-); \quad S_y = \frac{1}{2i}(S_+ - S_-) \tag{3.7}$$

with

$$L_\pm |m_L'\rangle = \sqrt{L(L + 1) - m_L(m_L \pm 1)} |m_L' \pm 1\rangle \tag{3.8}$$

$$S_\pm |m_S'\rangle = \sqrt{S(S + 1) - m_S(m_S \pm 1)} |m_S' \pm 1\rangle \tag{3.9}$$

Inserting Eq. 3.6 and 3.8 into Eq. 3.4 results in the final spin-orbit Hamiltonian:

$$H_{so} = \zeta \left[L_z S_z + \frac{1}{2}(L_+ S_- + L_- S_+) \right] \tag{3.10}$$

What is left is the definition of the operators L_i and S_i, with i being z, $+$, or $-$. Each of them is defined as a generalized Pauli matrix σ^N of order N, with N being $(2L + 1)$ or $(2S + 1)$ respectively (see Appendix A.3). To expand the dimension of these operators to a $(2L + 1) \times (2S + 1)$ Hilbert space we apply the Kronecker product $\otimes$. It is not commutative, and the order of the multiplication needs to be preserved. The operators L_i and S_i are therefore:

$$L_i = \sigma_i^{2L+1} \otimes \mathbb{I}^{2S+1}$$
$$S_i = \mathbb{I}^{2L+1} \otimes \sigma_i^{2S+1}$$

with $\mathbb{I}^M$ being the identity matrix of order M. Setting $\zeta = -336\,\mathrm{K}$ and diagonalizing the Hamiltonian results in the eigenvalues as shown in Fig. 3.6. The calculated eigen-

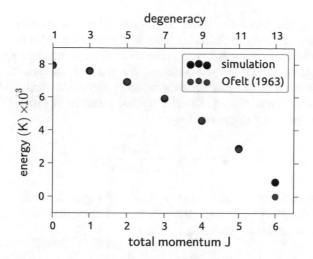

Fig. 3.6 This graph is obtained by calculating the eigenvalues of Eq. 3.10 and $\zeta = -336K$. The simulated values were shifted vertically to coincident with the values taken from [3]. As depicted the spin-orbit coupling lifts the degeneracy of the 49 states resulting in seven different multiplets with $J = 6$ as a new ground state.

Table 3.2 Energy splitting between the ground state (GS) $|J\rangle$ and excited state (ES) $|J - 1\rangle$ multiplets for pure LS coupling [2, 5]

Ion	Tb³⁺	Dy³⁺	Ho³⁺	Er³⁺	Tm³⁺	Yb³⁺
Elec. Conf.	$4f^8$	$4f^9$	$4f^{10}$	$4f^{11}$	$4f^{12}$	$4f^{13}$
GS	$^7F_6$	$^6H_{\frac{15}{2}}$	$^5I_8$	$^4I_{\frac{15}{2}}$	$^3H_6$	$^2F_{\frac{7}{2}}$
ES	$^7F_5$	$^6H_{\frac{13}{2}}$	$^5I_7$	$^4I_{\frac{13}{2}}$	$^3H_5$ (¹)	$^2F_{\frac{5}{2}}$
ΔE (K)	2900	4300	7300	9400	11900(¹)	14400

(¹) For Tm³⁺ $^3H_4$ lies below $^3H_5$ [2]

values and the experimentally obtained ones fit very well except for $J = 6$, where higher order perturbation theory is necessary. Nevertheless, a large energy splitting between the new ground state $J = 6$ and the new first excited state $J = 5$ of 2900 K [3, 4] is observed, making it possible to simplify the calculation of the magnetic properties by considering the 13 ground states only.

As shown in Table. 3.2 the large splitting between the ground state and the first excited state is a general property of rare earth ions and increases with the atomic number.

3.6 Ligand-Field Interaction

The ligand-field theory describes the electrostatic interaction between the coordination center of a complex and its ligands, leading to a modification of the electronic states of the former. Since the 4f shell of the lanthanides is situated inside the 5s and 5p shell, it is to a large part protected from its surrounding environment. However, the effect on the energy levels is still in the order of a few hundred Kelvin and acts as a perturbation on the spin-orbit coupling. We will start our considerations with a brief introduction into the ligand-field theory and apply this formalism to the terbium double-decker in the following. The electrostatic potential V_{lf} created by the ligand can be expressed in a very general way:

$$V_{lf}(r) = \int \frac{\rho(r')}{4\pi\epsilon_0 |r - r'|} d^3 R \tag{3.11}$$

where r is the position of the electron and $\rho(r')$ the charge density of the ligands. Since symmetry plays a very important role in this theory, we will express $1/|r - r'|$ in terms of spherical harmonics:

$$\frac{1}{|r - r'|} = \sum_{k=0}^{\infty} \frac{4\pi}{2k+1} \frac{r^k}{R^{k+1}} \sum_{q=-k}^{k} Y_k^q(\Theta, \Phi) Y_k^q(\theta, \phi) \tag{3.12}$$

where $Y_k^q(\Theta, \Phi)$ describes the position of the ligands, and $Y_k^q(\theta, \phi)$ describes the position of the electron. Therefore the ligand-field potential becomes:

$$V_{lf}(r) = \sum_{k=0}^{\infty} \sum_{q=-k}^{k} r^k \underbrace{\frac{4\pi}{2k+1} Y_k^q(\theta, \phi)}_{C_k^q} \underbrace{\int \frac{\rho(r')}{4\pi\epsilon_0 R^{k+1}} Y_k^q(\Theta, \Phi) d^3 R}_{A_k^q} \tag{3.13}$$

The term C_k^q is the so-called Racah tensor and depends only on the ligand position. The last term A_k^q is the geometrical coordination factor, which is a constant that can be determined experimentally. As V_{lf} can be treated as a perturbation to the spin-orbit ground-state multiplet, J remains a good quantum number, and the wave function Ψ can be written as $\Psi = |J, m_J\rangle$. It is very convenient to replace the operator C_k^q by the Stevens operators O_k^q, which are linear combinations of the total angular momentum operators and simplify the calculation in this basis [6]. Additional factors u_k (Stevens factors) account for the proper transformation [7]. The symmetry of the O_k^q is identical to the spherical harmonics Y_k^q, where $k - q$ is the number of nodes in the polar direction and q the number of nodes in the azimuthal direction with $-k \leq q \leq k$. The matrices for $q = 0$ have only diagonal elements, whereas for $q \neq 0$ off-diagonal elements occur, introducing a coupling between different states. The term O_0^0 has spherical symmetry and gives rise to a constant potential, which can be omitted. Furthermore, due to time reversal symmetry, all odd values of k vanish since they involve J_z to odd powers. It is sufficient to carry out the summation up to $k \leq 2J$ [8], with higher order terms being usually smaller than lower order terms. The ligand-field Hamiltonian H_{lf} is therefore:

Table 3.3 (a) The Stevens factors [7] and (b) the ligand-field parameters [10] for TbPc$_2$

(a)	u_2	u_4	u_6		
	$-\frac{1}{99}$	$\frac{2}{16335}$	$-\frac{1}{891891}$		
(b)	$A_2^0\langle r^2\rangle$	$A_4^0\langle r^4\rangle$	$A_4^4\langle r^4\rangle$	$A_6^0\langle r^6\rangle$	$A_6^4\langle r^6\rangle$
	595.7 K	-328.1 K	14.4 K	47.5 K	0 K

$$H_{lf} = \sum_{k=0}^{\infty} \sum_{q=-k}^{k} A_k^q \langle r^k\rangle u_k O_k^q (J_x, J_y.J_z) \qquad (3.14)$$

The matrix elements O_k^q of the Stevens operators are tabulated in [2] and Appendix B, and the terms $A_k^q\langle$ and $r^k\rangle$ can be determined experimentally using absorption spectra (Table 3.3).

Now we turn to the Hamiltonian for TbPc$_2$. Time reversal symmetry tells us that at zero magnetic field m_J and $-m_J$ are degenerate. Therefore, only even k values are allowed. Due to the decreasing weight of terms with higher order, we can limit the allowed k values to 2, 4, and 6. Furthermore, due to the local approximate C_4 symmetry of TbPc$_2$ the only remaining q values are $q = 0, 4$. With these considerations the final Hamiltonian of the TbPc$_2$ becomes [9]

$$H_{lf} = \langle r^2\rangle u_2 A_2^0 O_2^0 + \langle r^4\rangle u_4 \left(A_4^0 O_4^0 + A_4^4 O_4^4\right) + \langle r^6\rangle u_6 \left(A_6^0 O_6^0 + A_6^4 O_6^4\right) \quad (3.15)$$

with

The terms O_k^0, contain the operator J_z up to the power of k and are introducing a strong uni-axial anisotropy in z-direction. As a result, the degeneracy between $|J, m_j\rangle$ and $|J, m_j \pm 1\rangle$ is lifted, whereas due to the even powers of J_z the $|J, m_j\rangle$ and $|J, -m_j\rangle$ states remains degenerate. An exact numerical diagonalization of $H_{lf} + g_J \mu_B J_z B_z$ at different magnetic fields results in Fig. 3.7a. The ligand field induces an energy gap of a few hundred Kelvin between the ground state $|6, \pm6\rangle$ and the first excited state $|6, \pm5\rangle$. Therefore, already at liquid nitrogen temperatures, the magnetic properties of this complex are almost exclusively determined by the new ground state doublet $m_J = \pm6$. At room temperature the ground state population is still at 69 %. If we would replace the terbium ion by another rare earth ion like Dy^{3+}, Ho^{3+}, Er^{3+}, Tm^{3+}, or Yb^{3+}, this splitting would decrease as shown in Fig. 3.8 [11]. The terms O_4^4 and O_6^4 occur due to the slight misalignment between the two phthalocyanine ligands, which are not exactly rotated by 45°. Note that for an angle of 45 degrees the system would have a higher symmetry namely D_{4d}, resulting in the suppression of these two terms. Since the misalignment is only a few degrees, the geometrical coordination factor A_6^4 is still too tiny to be measured and can be omitted. The term O_4^4 contains the operators J_+^4 and J_-^4, which are mixing the ground state doublet and lift their degeneracy by $\Delta \simeq 1$ μK (see Fig. 3.7b). This so-called avoided level crossing gives rise to zero field tunneling of the magnetization, which will be explained in Sect. 3.8.1.

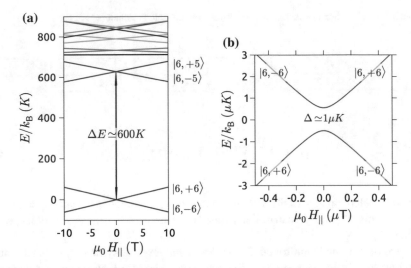

Fig. 3.7 Zeeman diagram of the TbPc₂. **a** The ligand field splits the ground state (*red*) and first excited state (*blue*) by around 600 K, leaving only two spin degrees of freedom at low temperature, which makes the molecule an ideal two level quantum system. Higher order excited states are $|6, 0\rangle$ (*black*), $|6, \pm1\rangle$ (*green*), $|6, \pm2\rangle$ (*orange*), $|6, \pm3\rangle$ (*grey*) and $|6, \pm4\rangle$ (*purple*). **b** Additional terms in the ligand field Hamiltonian (A_4^4, A_6^4) lift the degeneracy of the ground state doublet by $\Delta \simeq 1\,\mu K$ and introduce an avoided level crossing in the Zeeman diagram.

Fig. 3.8 Crystal field splitting of the bis-phthalocyaninato complex with different rare earth atoms as coordination centers (adapted from [11]).

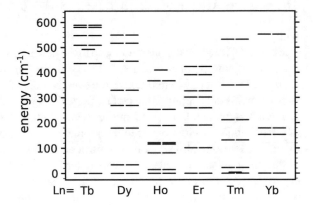

3.7   Hyperfine Interaction

The nucleus of the terbium ion has, besides its electrical charge, also an inherent angular moment $I = 3/2$, resulting in an additional magnetic dipole moment:

$$\mu_I = g_I \mu_N I \tag{3.16}$$

with $g_I = 1.354$ [12] and μ_N the nuclear magneton. Similar to the spin-orbit interaction, this magnetic moment interacts via dipole coupling with the magnetic moment μ_J created by total angular momentum J. The Hamiltonian accounting for this interaction is formulated as:

$$H_{dip} = A\,\boldsymbol{I}\boldsymbol{J} \tag{3.17}$$

$$\boldsymbol{I}\boldsymbol{J} = I_z J_z + \frac{1}{2}(I_+ J_- + I_- J_+) \tag{3.18}$$

with A being the hyperfine constant. To obtain Eq. 3.18 we use the same transformation as in Eq. 3.6.

In addition, the nuclear spin possesses an electric quadrupole moment which makes it sensitive to electric field inhomogeneities, such as produced by the electrons in the 4f orbitals. The Hamiltonian which accounts for this interaction can be written as:

$$H_{quad} = P\,(\boldsymbol{I}\boldsymbol{J})^2 \tag{3.19}$$

$$(\boldsymbol{I}\boldsymbol{J})^2 = (I_z J_z + \frac{1}{2}(I_+ J_- + I_- J_+))^2 \tag{3.20}$$

with P being the hyperfine quadrupole constant. The hyperfine Hamiltonian is now simply the sum of the magnetic dipole interaction and the electric quadrupole contribution.

$$H_{hf} = A\,\boldsymbol{I}\boldsymbol{J} + P\,(\boldsymbol{I}\boldsymbol{J})^2 \tag{3.21}$$

For the terbium ion the two parameters A and P are given in Table. 3.4.

By diagonalizing the full Hamiltonian

$$H = H_{lf} + H_{hf} + H_Z \tag{3.22}$$

at different magnetic fields and plotting the eight lowest lying eigenvalues, we obtain Fig. 3.9a. Due to the hyperfine interaction each electronic ground state is split in to four. The lines with a positive (negative) slope correspond to the electronic spin $|+6\rangle$ ($|-6\rangle$) and lines with the same color (blue, green, red, black) to the same nuclear spin state ($|+3/2\rangle, |+1/2\rangle, |-1/2\rangle, |-3/2\rangle$). The splittings of the electronic levels are unequal due to the quadrupole contribution of the hyperfine interaction and calculated as 2.(5) GHz, 3.(1) GHz and 3.(7) GHz as depicted in Fig. 3.9b. Moreover, the anticrossing, which was formerly at $B = 0$ T, is now split into four anticrossings,

Table 3.4 Hyperfine constant A and the quadrupole parameter P for the terbium ion according to Ishikawa et al. [10].

A	P
24.9 mK	0.4 mK

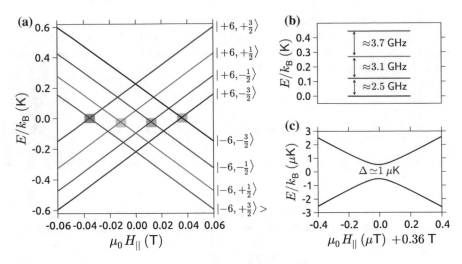

Fig. 3.9 a Zeeman diagram of the TbPc$_2$. The *colored rectangles* indicate avoided level crossing between two states of opposite electronic spin and identical nuclear spin. **b** Energy spacing between the different nuclear spin states. **c** Magnification of the avoided level crossing between $|6, \frac{3}{2}\rangle$ and $|-6, \frac{3}{2}\rangle$.

one for each nuclear spin state (colored rectangles in Fig. 3.9a). The energy gap at each anticrossing remains about 1 μK (Fig. 3.9c).

3.8 Magnetization Reversal

Changing the external magnetic field parallel to the easy axis of the TbPc$_2$ allows for the reversal of the molecule's magnetic moment. Hence, when sweeping the magnetic field periodically between positive and negative values we can measure a hysteresis loop as depicted in Fig. 3.10a. Is shows that the magnetization reverses in a step-like shape at small magnetic fields, followed by a continuous reversal at larger magnetic fields. The hysteresis shape can be understood by considering two completely different reversal mechanisms: a direct relaxation, dominating at larger magnetic fields; and the quantum tunneling of magnetization, dominating at small magnetic fields.

3.8.1 Quantum Tunneling of Magnetization

The quantum tunneling of magnetization (QTM) is a tunnel transition between two different spin states $|S, m_s\rangle$ and $|S, m'_s\rangle$. It requires a finite overlap of the two wavefunctions, which is caused by off-diagonal terms in the Hamiltonian. Since these

Fig. 3.10 **a** Normalized
hysteresis loop of a single
TbPc$_2$ single molecule
magnet obtained by
integration of 1000 field
sweeps. Adapted from [13].
b Zeeman diagram
calculated by diagonalizing
Eq. 3.22. The steps in the
hysteresis loop of (**a**)
coincide with the avoided
level crossings and are
caused by the quantum
tunneling of magnetization
(QTM). The remaining
magnetization reversal of (**a**)
can be explained by direct
transitions (DT) from the
excited state into the ground
state involving the creation
of a phonon.

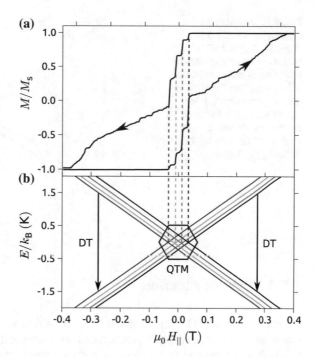

terms are usually small compared to the diagonal terms, the overlap is negligible except for those longitudinal magnetic fields, where the diagonal terms in the
Hamiltonian start to vanish. The consequence is the formation of an avoided energy
level crossing at those magnetic fields (see Fig. 3.11). When sweeping the longitudinal magnetic field over this anticrossing (see Fig. 3.11) the spin can tunnel from
the $|S, m_{\mathrm{s}}\rangle$ into $|S, m_{\mathrm{s}}'\rangle$ state with the probability P given by the Landau-Zener
formula [14, 15]:

$$P_{m,m'} = 1 - exp\left(-\frac{\pi \Delta_{m,m'}}{2\hbar g\mu_{\mathrm{B}}|m - m'|\mu_0 dH_{||}/dt}\right) \tag{3.23}$$

Formula 3.23 states that the transition probability increases exponentially with the
level splitting Δ and decreases exponentially with the sweep-rate $\mu_0 dH_{||}/dt$ of the
longitudinal magnetic field.

As described in Sect. 3.7 the TbPc$_2$ possesses four of these avoided level crossings.
This results in four distinct steps at small magnetic field in Fig. 3.10a, which can
be used as a fingerprint to identify the single-molecule magnet, as it was shown by
Vincent et al. [16].

Fig. 3.11 Avoided level
crossing between the two
states $|S, m_\mathrm{s}\rangle$ and $|S, m'_\mathrm{s}\rangle$,
which leads to the quantum
tunneling of magnetization.
While sweeping the parallel
field over the anticrossing
the probability P to tunnel
from one state into the other
is given by the Landau-Zener
formula (Eq. 3.23).

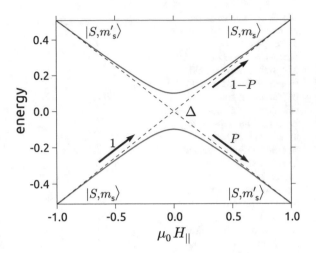

3.8.2 Direct Transtions

In addition to the QTM, the magnetic moment of the molecule can reverse in a direct
transition. This is an inelastic process and involves the creation and/or annihilation
of phonons to account for the energy and momentum conservation. Therefore, this
process is often referred to as phonon assisted or spin-lattice relaxation.

Depending on the temperature we can distinguish between three types of relaxation
processes. At low temperature the spin of an SMM is most likely reversed in a
direct relaxation process under the emission of one phonon to the thermal bath (see
Fig. 3.12a). This process becomes more likely at higher magnetic fields and scales
with $(\mu_0 H)^3$. Increasing the temperature allows for a two phonon relaxation process.
Therein, the molecule is excited into the state $|e\rangle$ while absorbing a phonon of energy
$\hbar\omega_1$ and subsequently relaxes into the ground state via the emission of a phonon of
energy $\hbar\omega_2$. Depending on whether the excited state is a real or virtual state, we
distinguish between the Orbach process (see Fig. 3.12b) or the Raman process (see

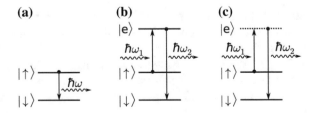

Fig. 3.12 Phonon assisted relaxation. **a** Direct relaxation into the ground state under the emission
of a phonon with energy $\hbar\omega$. **b** The two phonon Orbach process involves the absorption of a phonon
of energy $\hbar\omega_1$, exciting the molecule into the state $|e\rangle$, and a subsequent emission of another phonon
of energy $\hbar\omega_2$, relaxing the molecule into its ground state. **c** The two phonon Raman process is
similar to the Orbach process, however, the excited state $|e\rangle$ is a virtual state.

Fig. 3.12a). The Orbach process shows an exponential temperature dependence $\frac{1}{\tau} \propto exp(\Delta/k_B T)$, whereas the Raman process has a polynomial temperature dependence $\frac{1}{\tau} \propto (k_B T)^7$, with τ being the relaxation time, $\Delta = \hbar\omega_2 - \hbar\omega_1$, and T the temperature.

References

1. H. Lueken, *Course of lectures on magnetism of lanthanide ions under varying ligand and magnetic fields* (Institute of Inorganic Chemistry, RWTH Aachen, 2008)
2. A. Abragam, B. Bleaney, *Electron Paramagnetic Resonance of Transition Ions (Oxford Classic Texts in the Physical Sciences)* (Oxford University Press, USA, 2012). ISBN 0199651523
3. G.S. Ofelt, Structure of the f6 Configuration with Application to Rare-Earth Ions. J. Chem. Phys. **38**, 2171 (1963)
4. K.S. Thomas, S. Singh, G.H. Dieke, Energy Levels of Tb3+ in LaCl3 and Other Chlorides. J. Chem. Phys. **38**, 2180 (1963)
5. K. Binnemans, R. Van Deun, C. Görller-Walrand, J. Adam, Spectroscopic properties of trivalent lanthanide ions in fluorophosphate glasses. J. Non-Cryst. Solids **238**, 11–29 (1998)
6. D. Smith, J.H.M. Thornley, The use of operator equivalents. Proc. Phys. Soc. **89**, 779–781 (1966)
7. K.W.H. Stevens, Matrix elements and operator equivalents connected with the magnetic properties of rare earth ions. Proc. Phys. Soc. Sect. A **65**, 209–215 (1952)
8. D. Gatteschi, R. Sessoli, J. Villain, *Molecular Nanomagnets* (Oxford University Press, USA, 2006). ISBN 0198567537
9. C. Gröller-Walrand, K. Binnemans, *Handbook on the Physics and Chemistry of Rare Earths.* Elsevier Amsterdam, 23 edition (1996). (ISBN 9780444825070)
10. N. Ishikawa, M. Sugita, W. Wernsdorfer, Quantum tunneling of magnetization in lanthanide single-molecule magnets: bis(phthalocyaninato)terbium and bis(phthalocyaninato)dysprosium anions. Angewandte Chemie (International ed. in English) **44**, 2931–2935 (2005)
11. N. Ishikawa, M. Sugita, T. Okubo, N. Tanaka, T. Iino, Y. Kaizu, Determination of ligand-field parameters and f-electronic structures of double-decker bis(phthalocyaninato)lanthanide complexes. Inorg. Chem. **42**, 2440–2446 (2003)
12. J.M. Baker, J.R. Chadwick, G. Garton, J.P. Hurrell, E.p.r. and Endor of Tb Formula in Thoria. Proc. R. Soc. A: Math. Phys. Eng. Sci. **286**, 352–365 (1965)
13. R. Vincent, S. Klyatskaya, M. Ruben, W. Wernsdorfer, F. Balestro, Electronic read-out of a single nuclear spin using a molecular spin transistor. Nature **488**, 357–360 (2012)
14. L. Landau, Zur Theorie der Energieubertragung II. Phys. Sov. Union **2**, 46–51 (1932)
15. C. Zener, Non-adiabatic crossing of energy levels. Proc. Roy. Soc. A: Math. Phys. Eng. Sci. **137**, 696–702 (1932)
16. R. Vincent, S. Klyatskaya, M. Ruben, W. Wernsdorfer, F. Balestro, Electronic read-out of a single nuclear spin using a molecular spin transistor. Nature **488**, 357–360 (2012)

Chapter 4
Experimental Details

4.1 Overview Setup

During my thesis, my work was dedicated to the study of single-molecule magnet based transistors in order to perform a coherent quantum manipulation and a non-destructive read-out of a single nuclear spin. Towards this goal, I designed an experimental setup to perform ultra low noise electrical measurements at very low temperature (40 mK), under the influence of fast sweeping 3D magnetic fields and RF electromagnetic fields. An overview of the entire setup is presented in Fig. 4.1. In interaction with the different technical supports of the Néel Institut, I fabricated and tested the different parts of the setup, which were designed to fulfill the diverse experimental constraints of this experiment.

The molecular spin transistor is a three terminal device, consisting of a single-molecule magnet (TbPc$_2$), which is electrically coupled to source, drain, and gate electrodes. In order cool down the device to very low temperatures, it was mounted onto the cold finger of a dilution refrigerator whose base temperature is about 40 mK.

The transistor was microbonded on a specially designed chip carrier consisting of a 50 Ω broadband waveguide and 24 DC strip lines. To avoid 4 K radiation, this chip carrier was encapsulated in a fixed radiation shield anchorage to the mixing chamber. A large sweep-rate, three-dimensional vector magnet, surrounding the chip carrier, was developed to control and read-out the anisotropic electronic moment carried by the single molecular magnet. Electrical connections of the spin transistor to the outside world were established via low temperature pi-filters (1 MHz–1 GHz) and home-made Eccosorb filters (from 1 GHz). Subsequently, at room temperature, the signal can be amplified by two different current-voltage converters. One was dedicated to the electromigration procedure, while the other one was designed for very sensitive low current measurements. They were directly connected to the cryostat to minimize the electro-magnetic and electro-mechanic noise pick up. Additionally, we used room temperature low pass filters and voltage dividers on the bias and gate voltages wires to reduce the noise which was send to the sample. All this room

© Springer International Publishing Switzerland 2016
S. Thiele, *Read-Out and Coherent Manipulation of an Isolated Nuclear Spin*,
Springer Theses, DOI 10.1007/978-3-319-24058-9_4

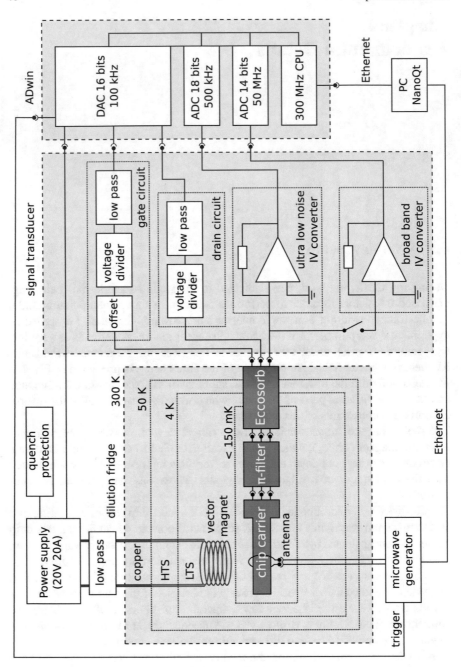

Fig. 4.1 Schematic of the experimental setup

temperature electronics is integrated into a doubly shielded box, which we refer to as the signal transducer.

Finally, the signal transducer is controlled via an independent real time digital-analog converter (ADwin). The latter drives also the 3D magnetic field and triggers the microwave pulse generator, which guarantees a synchronized operation of all devices.

Fig. 4.2 a Picture of the experimental setup. The computer (**f**) controls the microwave source (**g**) and the ADwin real time data acquisition unit (**c**). The latter, in turn, operates the signal transducer (**d**) and the power supplies of the vector magnet, which are situated in the basement (**e**) via a remote terminal (**b**). The three LEDs in (**b**) indicate if the coils are operating. In case of a quench, the corresponding power supply is shut down automatically, and the LEDs will turn off. Further details of the individual parts of the experiment are explained in the text

The ADwin is connected over Ethernet with a standard PC, which interfaces the unit using a home-made software called NanoQt.

A picture of the entire setup is shown in Fig. 4.2, whereas a more detailed description of all experimental parts is given in the following sections.

4.2 Dilution Refrigerator

To explore of the quantum world of a molecular spin transistor a low temperature environment is required, which makes the use of dilution refrigerators (DR) indispensable. Among many different concepts, we chose to work with an inverted DR, which combines a fast cool down (about 3 h) with a spacious low temperature stage. The basic working principle of this DR will be explained in the following paragraphs. The schematic in Fig. 4.3 shows that cryostat consists of six different thermal stages, each encapsulated by another with higher temperature. Vacuum isolates one level from another so that each stage functions as a radiation shields for the next inner lying. To cool down the cryostat, two independent cooling circuits operate simultaneously. The secondary open cycle cooling circuit replaces the liquid ^{4}He bath of conventional cryostats (green circuit in Fig. 4.3). It operate with liquid ^{4}He, which is injected from a Dewar underneath the DR into the so called 4 K box. Since the Dewar is slightly over pressured, a sufficiently large ^{4}He circulation is established to guarantee a steady state operation. An additional pump inside the circuit is only needed during the cool down from room temperature, since high cooling power and hence high flow rates are necessary. The liquid helium inside the 4 K box is used to cool the 4 K stage, whereas the vapor created by the boiling liquid ^{4}He is ejected into a spiral counter-flow heat-exchanger. While leaving the cryostat, it gradually cools down the primary cooling circuit as well as the 20 K and the 100 K stages.

The primary cooling circuit is a closed cycle cooling circuit, containing a mixture of ^{3}He and ^{4}He. It is subdivided into a fast and slow injection (blue and red circuit in Fig. 4.3), both entering the DR via the counter-flow heat exchanger. Due to the cooling power extracted from the secondary circuit, the gas is gradually cooled down to 4.2 K. Afterward ,the fast injection is directly thermalized onto the 1 K stage and leaves the cryostat via the mixing chamber, the discrete exchangers, and the still. It has a larger cross section than the slow injection and is used to precool the colder parts of the cryostat to 4.2 K during the cool down from room temperature.

The slow injection on the other hand is responsible for the condensation of the mixture followed by the steady state operation. In order to condense the mixture, an external compressor pressurizes the gas to 4 bar before injecting it into the spiral heat exchanger of the cryostat. Leaving the latter at a temperature of 4.2 K, it passes a second heat exchanger, which is terminated by a flow impedance. The resulting pressure gradient leads to a Joule-Thomson expansion and lowers the temperature of the gas by ≈ 2 K before entering the still. After having passed the latter, the mixture traverses a set of continuous and discrete heat exchangers before being injected into the mixing chamber (Fig. 4.4).

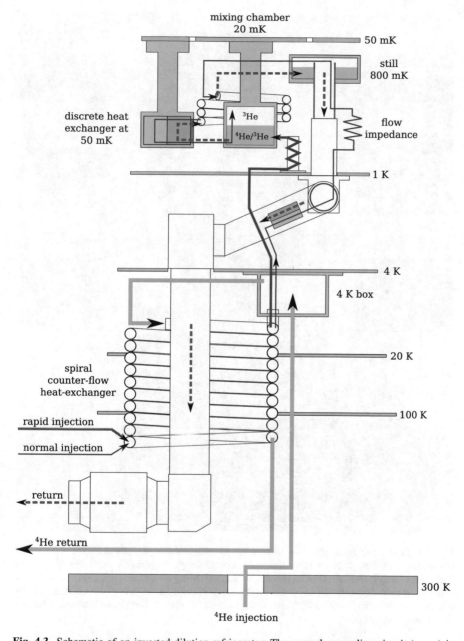

Fig. 4.3 Schematic of an inverted dilution refrigerator. The secondary cooling circuit (*green*) is precooling the primary circuit consisting of the normal injection (*red*) and rapid injection (*blue*). The latter is only used during the cool down from room temperature. During the steady state operation, ^{3}He is injected via the normal injection into the ^{3}He rich phase of mixing chamber and extracted from the diluted phase (*violet*)

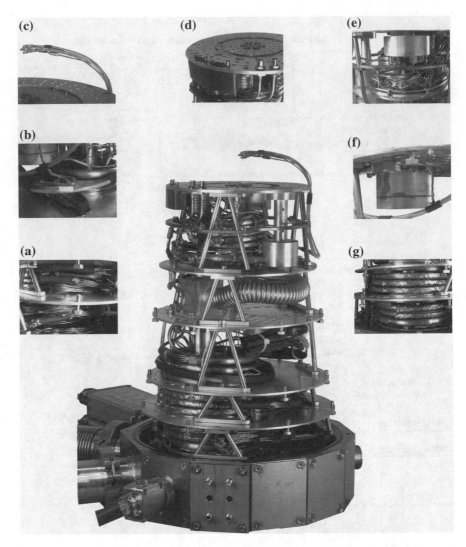

Fig. 4.4 *Center* Picture of the fully wired dilution refrigerator. **a–c** Current leads for the super-conducting vector magnet consisting of copper (**a**), high temperature superconducting (**b**), and low temperature superconducting cables (**c**). **d** Cold stage showing the DC and microwave connectors. The sample holder (not shown) is situated in the *center* of the cold stage. **e–g** Important parts of the primary and secondary cooling circuit showing the still (**e**), the 4 K box (**f**), and the spiral counter flow heat exchanger (**g**)

External pumps are decreasing the pressure inside the mixing chamber below 0.1 mbar, allowing for another adiabatic expansion, which results in the conden-sation of the mixture. The cold gas evaporating from the liquid is being pumped out through the numerous heat-exchangers cooling down the incoming mixture. Hence,

Fig. 4.5 Phase diagram of liquid ^{3}He and ^{4}He mixtures at saturated vapour pressure taken from [1]. Below the critical temperature of 867 mK, the mixture separates into two phases, a ^{3}He rich and a ^{3}He diluted phase

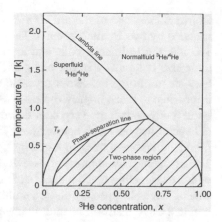

more and more gas condenses, gradually filling up every part from the mixing chamber to the still with liquid. At a temperature of around 800 mK, a phase separation into a lighter ^{3}He rich phase and heavier ^{3}He dilute phase is taking place inside the mixing chamber. The exact ration of ^{3}He/^{4}He in each phase depends on temperature and is shown in Fig. 4.5.

The diluted phase expands from the bottom of the mixing chamber to the still. It contains mainly super-fluid ^{4}He, which can be viewed as inert and noninteracting with the ^{3}He. Nevertheless, the vapor inside the still contains, despite the high concentration of ^{4}He, 97 % of ^{3}He due to its low boiling point. By pumping on the still and re-injecting the gas in the ^{3}He rich phase, a ^{3}He circulation is established. In order to maintain the equilibrium concentration, ^{3}He from the rich phase is pushed into the diluted phase. This is an endothermic process, providing the cooling power to cool down to mK temperatures. This process can also be viewed as an evaporation of liquid ^{3}He from the rich into the diluted phase since the ^{4}He, which requires heat and continues even to the lowest temperatures since the concentration of ^{3}He in the diluted phase remains finite. The base temperature of the cryostat is only determined by residual heat leaks and remains usually above 10 mK for most of the DRs. A picture of the fully mounted dilution refrigerator is shown in Fig. 4.4.

4.3 3D Vector Magnet

The observation and manipulation of a single-molecule magnet (SMM), which is the centerpiece of a molecular spin transistor, demands external magnetic fields in arbitrary directions. A way to create such three dimensional fields comprises three coils mounted perpendicular to each other like the axes of a coordinate system. The orientation and magnitude of the magnetic field is controlled by adjusting the current through each coil so that the resulting field is simply the vector sum of three respective fields.

Conventional state of the art 3D vector magnets consist of a cylindrical coil surrounded by two Helmholtz coils. They are capable of creating a magnetic field of 1 T with the Helmholtz coils and around 2 T with the cylindrical coils. However, their size is typically in the order of $200 \times 200 \times 200$ mm. Despite the fact that they are not fitting inside our cryostat, they have a very high inductance making it impossible to reach high sweep rates. Furthermore, their huge heat capacity would be severely retarding every cool down. Therefore, we aimed to build very small 3D vector magnets with approximately the same magnetic field specifications. The fabrication process was supported by Yves Deschanels from the Institute Néel.

The cryogenic environment of the DR allows for the use of superconducting wires, which are creating much higher fields than conventional copper wires. Among the several available types, we chose a multifilament NbTi superconducting wire embedded in a CuNi matrix. The multifilament layout diminishes flux jumps and reduces the total amount of vortices, leading to higher stability and smaller remanence. The NbTi superconducting core is known to be less fragile than the Nb_3Sn core, which was important during the fabrication process. The CuNi matrix was chosen because of smaller Eddy-currents compared to a pure Cu matrix, hence, allowing for higher sweep rates.

Since the DR operates in vacuum, the maximum current per coil was fixed to 20 A for field pulses and 10 A for steady state operation as safety precautions. Looking up the different specifications of available SC wires, we found the low current SC wire from SUPERCON Inc. with an outer diameter of 152 μm and 18 NbTi filaments as most suited for our purpose.

A first design study was then carried out to optimize the central-cylindrical coil, also referred to as the z-coil, using the COMSOL Multiphysics software. The field at the sample, situated inside the z-coil, should be around 1 T at 10 A in order to be comparable with commercial state of the art electromagnets. The inner diameter of the coil was set to 6 mm, thus, being still large enough to insert the chip carrier with the sample later on. Given the current, the wire diameter, and an ideality factor of coil of 98 %, the parameters remaining for optimization were the coil length L and the width W of the accumulated layers. The calculated magnetic field in the parameter space of L and W is displayed in Fig. 4.6. In order to maximize the field of the other two coils, W must be as small as possible. As shown in Fig. 4.6 the optimal dimensions were found to be $W = 2$ mm and $L = 25$ mm.

With the above mentioned dimensions of the z-coil ($W = 2$ mm and $L = 25$ mm), we calculated the spacial magnetic field distribution. The result of this simulation (Fig. 4.7) shows an almost uniform field distribution within a radius of 3 mm around the center, which is about the size of our sample.

Having set the dimensions of the z-coil, we started the design-study of the x and y split-pair magnets. Their separation of 10 mm is given by the outer diameter of the z-coil. In order to reach fields of around 1 T at 10 A, with a coil separation of 10 mm, we developed a new design concept. In a first approach, we replaced the standard cylindrical Helmholtz coils by conically shaped coils, thus, increasing the volume share at equal dimensions. Fixing the smaller diameter of the cone to 10 mm, we end up with two variable parameters, namely, the inner diameter D and the coil thickness

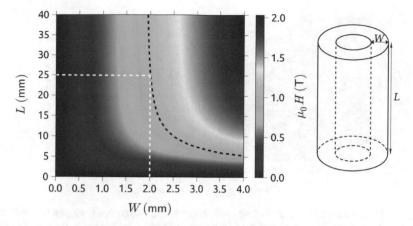

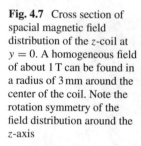

Fig. 4.6 Magnitude of the magnetic field in the center of the z-coil as a function of the coil length L and the accumulated width W of the wire layers. The *black dashed line* is the isofield line of 1 T. In order to reach 1 T at 10 A the minimum width needs to be 2 mm resulting in a length of 25 mm

Fig. 4.7 Cross section of spacial magnetic field distribution of the z-coil at $y = 0$. A homogeneous field of about 1 T can be found in a radius of 3 mm around the center of the coil. Note the rotation symmetry of the field distribution around the z-axis

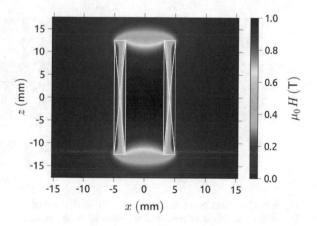

L. In order to find the optimal parameters, a second design study was carried out. The calculated magnetic field in the parameter space of D and L is shown in Fig. 4.8. For engineering reasons the conical shape needed to be approximated by a step like shape. In the first iteration, we introduced only one step in each coil. The position of this step was subsequently optimized, by keeping the above mentioned thickness and inner diameter, in order to obtain 1 T at 10 A in the center of the vector magnet. The final result of the shape and magnetic field magnitude for the x and y coil is shown in Fig. 4.9a, b respectively. Notice that the highest field of the split coils is in the center of the inner wall and is much higher than the field at the center of the vector magnet.

A picture of the fully mounted vector magnet is shown in Fig. 4.10. The first tests were carried out in liquid helium. We measured the maximum magnetic field as well

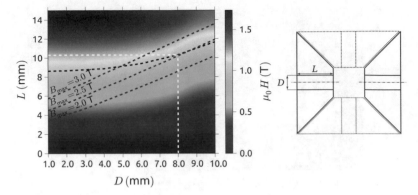

Fig. 4.8 Magnitude of the magnetic field in the center of the two conically shaped split coils as a function of the inner diameter D and the thickness L. The *black dashed line* is the isofield line of 1 T. The *red line* correspond to the maximum field at the inner wall of the coils. In order to reduce that maximum field, we set the dimensions to $D = 8$ mm and $L = 10$ mm

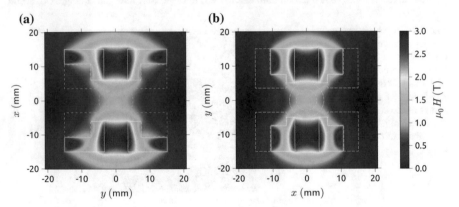

Fig. 4.9 Cross section of the magnetic field distribution of the x-coil (**a**) and y-coil (**b**) at $z = 0$. The contour of the respective coil is shown in *white*, whereas the contours of the other two coils are drawn as a *grey dotted line*. The generated field in the center of the vector magnet is about 1 T at 10 A for each coil and the maximal field is around 2.7 T. Note that the field distribution of the x- and y-coil has an axial symmetry around the x- and y-axis respectively

as the maximum sweep rate, the coils could resist before quenching. The results are shown in Table 4.1.

Each coil was able to produce a field of 1 T when operated alone, however, the x and y coil generated this field only at 11 and 12.5 A respectively. When operating all three coils simultaneously, the maximum field was limited to 0.9 T due the mutual interaction.

In the vacuum environment of the dilution refrigerator, the maximum field is reduced by $\approx 20\%$ and the maximum sweep speed by a factor of 5. This is caused by the slightly higher temperature of 4.4 K instead of 4.2 K and less efficient thermalization

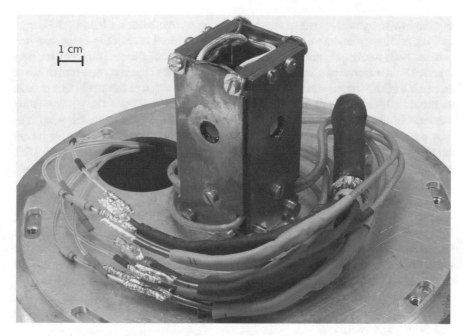

Fig. 4.10 Picture of the fully mounted vector magnet

Table 4.1 Benchmarks of the three different coils, which were immersed in liquid helium

	x-coil	y-coil	z-coil
I_{max}	15 A	14 A	18 A
B @ 10 A	0.9 T	0.8 T	1.1 T
$(dB/dt)_{max}$	≈1 T/s	≈1 T/s	>10 T/s

of the coils. In liquid helium, any generated heat is directly mediated to the liquid helium bath, whereas in the vacuum environment of the DR, the heat has to diffuse to the copper thermalization of the coils, which creates a bottleneck in the thermal transport.

4.4 Current Leads

The current leads are the electrical link between the superconducting vector magnet and the room temperature connections outside the cryostat. To guarantee a stable operation, an equilibrium between the wire material, the diameter, and the length had to be found. Ideally, the material should be a very good electrical conductor but a very bad thermal conductor.

For the low temperature part of the cryostat, i.e. at temperatures below 77 K, high temperature superconductors (HTS) were used as current leads. Since superconductors are both perfect electrical conductors and very poor thermal conductors, they represent the material of choice. The HTS we used in the cryostat consisted of silver coated YBaCuO straps with a T_c of 90 K. The cross section of the straps was chosen to sustain 40 A at 77 K, thus, leaving a safety coefficient of two. The silver coating was needed to achieve a homogeneous temperature of the HTS along the strap. The drawback of these kind of superconductors is their fragility. Therefore, we terminated the HTS straps with low temperature superconductors of the NbTi type. This simplifies the soldering and unsoldering of the current leads from the vector magnet, which was necessary every time the sample is changed.

The high temperature part of the current leads was made of copper wires. Since the resistivity and thermal conductivity of copper varies with temperature, a design study was carried out to determine the optimal geometry. Yet, a too large diameter results in a thermal shortcut between stages of different temperatures, whereas a too small diameter could destroy the leads due to Joule heating. The same considerations can be made for the wire length L, since the heat conduction is proportional to $1/L$. Therefore, a very short wire will transmit a lot of heat into the dilution fridge, whereas a very long cable might not be able to remove the energy produced by Joule heating and the wire possibly melts.

To work out this optimization problem, the one dimensional heat equation was solved with the experimental boundary conditions. It is a inhomogeneous partial differential equation and given as [2]:

$$\frac{dT}{dt} - \frac{1}{c\rho}\frac{d}{dx}\left(\kappa\left(T\right)\frac{dT}{dx}\right) = \frac{\dot{q}}{c\rho} \tag{4.4.1}$$

$$\frac{dT}{dt} - \frac{1}{c\rho}\frac{d\kappa\left(T\right)}{dx}\frac{dT}{dx} - \frac{\kappa T}{c\rho}\frac{d^2T}{d^2x} = \frac{\dot{q}}{c\rho} \tag{4.4.2}$$

where κ is the thermal conductivity, c the specific heat capacity, ρ the density, T the temperature, and $\dot{q}$ the heating power per unit volume. The effect of the black body radiation was neglected since it is much smaller than the other parts of the equation within the temperature range of the experiment. The term on the right hand side is the source term and corresponds to the energy injected into the system. This energy is partly used to heat the wire (that is where the $\frac{dT}{dt}$ comes from) and is partly transported away, which gives rise to the $\frac{\kappa}{c\rho}\frac{d^2T}{dx^2}$ term. For a metal wire the power $\dot{q}$ due to Joule heating is given by:

$$\dot{q} = \frac{U \cdot I}{A \cdot L} = \frac{I^2 R}{A \cdot L} = \frac{I^2}{A^2}\frac{1}{\sigma\left(T\right)} \tag{4.4.3}$$

where I is the applied current, A is the cross section area, L the length of the wire, and σ the electrical conductivity of the metal.

Table 4.2 Results of the numerical optimization of the heat equation for the optimal wire length L and the optimal wire diameter D in the different temperature regions

	L (cm)	A (mm^2)
300–200 K	30	1.5
200–100 K	25	0.75
< 100 K	25	0.5

The electrical conductivity $\sigma(T)$ of copper in the temperature range of 50 to 300 K can be modeled as [3]:

$$\frac{1}{\sigma(T)} = -3.204 \cdot 10^{-9} + 6.855 \cdot 10^{-11}\,T \quad [\Omega\text{m}] \tag{4.4.4}$$

Another important parameter is the thermal conductivity $\kappa(T)$ of copper. In the range from 100 to 300 K it can be fitted by [4]:

$$\kappa(T) = 886 - 7.462 \cdot T + 0.045 \cdot T^2 - 1.2331 \cdot 10^{-4} \cdot T^3 + 1.267 \cdot 10^{-7} \cdot T^4 \tag{4.4.5}$$

and from 50 to 100 K by [4]:

$$\kappa(T) = 7051 - 277.4T + 4.69T^2 - 0.0368T^3 + 1.106 \cdot 10^{-4}T^4 \tag{4.4.6}$$

Since the electrical and thermal conductivity of copper increases for decreasing temperature, the diameter of the leads needs to be decreased to minimize heat leaks. For practical reasons the diameter reduction is done at two temperatures: 200 and 100 K. Using Eq. 4.4.2 in combination with Eqs. 4.4.3–4.4.6, the optimal parameters for the wire length L and the diameter D at the different temperature ranges were calculated. The optimization parameters were such that the created heat leak should be less than 1.7 W, which corresponds to $\approx$10% of the cooling power of the primary circuit at a ^{4}He flow rate of 3.6 l/min; and that the temperature increase during a steady state operation at 10 A remains smaller than 10 %. The results for the three different temperature regions are tabulated in Table 4.2.

The values given in Table 4.2 do not include the size of the thermalizations, which were chosen to be 20 cm at 200 K, 12 cm at 100K, and 16 cm at 50 K. They were realized by gluing copper litz wires onto the current leads with a mixture of araldite and silver powder.

In order to operate the vector magnet, six of these current leads (two for each coil) were fabricated. After having them installed together with the HTS, we tested the ensemble at 10 A per lead. During the test, the temperature of the outermost stage in the cryostat increased by about 50 K, whereas the temperature-increase of the 20 K stage was already below 1 K, so that the stable operation of the dilution fridge was guaranteed.

4.5   Sample Holder

The sample holder is the link between the sample and the cryostat. It consists of two parts, an exchangeable chip carrier and a fixed radiation shield, which is in direct contact with the mixing chamber of the cryostat. It is needed to block the 4 K radiation of the vector magnet and keeps the sample at mK temperatures. The sample holder was designed to have an independent vacuum, which protects the sample when heating up the cryostat to room temperature. A picture of the radiation shield is shown in Fig. 4.11.

The chip carrier was designed to have 24 DC strip lines and one 50 Ω matched broadband waveguide. It is connected via a 36 Pin PCI Express connector to the radiation shield, which, when it is closed, encapsulates the chip carrier. The chip carrier itself is made out of six copper/insulator layers, which are shown in Fig. 4.12.

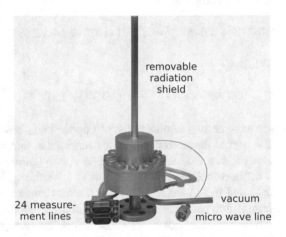

Fig. 4.11 Radiation shield with a independent vacuum, a feed through of 24 measurement lines and one mircowave line

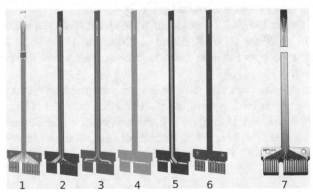

Fig. 4.12 (*1–6*) Layout of the chip carrier consisting of six independent layers. The *top three layers* contain the 24 DC strip lines and *three bottom layers* 50 Ω matched waveguide. (*7*) Picture of the sample holder

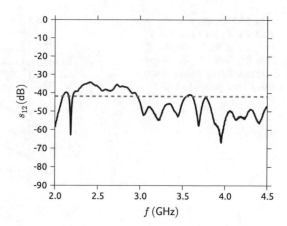

Fig. 4.13 Microwave transmission s_{12} measured from the SMA connector to the end of the waveguide using an Agilent E8362C vector network analyzer

The three top most layers contain the 24 DC strip lines, layer four, five, and six are used for the 50 Ω matched waveguide. The DC lines are soldered to two 12 pin Cannon connectors and the waveguide to a SMA terminated microcoax. Despite the 50 Ω matching the transmission s_{12} measured from the SMA connector to the end of the waveguide on the top layer is around -40 dBm (see Fig. 4.13). A large part of the attenuation is probably coming from reflections at the PCI Express interface. Insertion losses of the microcoax are about 10.5 dBm/meter at 1 GHz and have only a minor influence.

4.6 Filter

Most experiments exploring the quantum nature of matter are sensitive to external noise sources, which, if not properly attenuated, decrease the coherence time of a quantum state drastically. In general, there are three main noise source interfering with the experiment.

The first one is the noise generated by electro-magnetic radiation. It is produced by any wireless communication system and is in the order of a few Hz to a few GHz, e.g. Wifi, mobile phones, television, GPS, etc., or by improperly shielded power sources like any switching power supply or transformer.

The second noise source is the Johnson-Nyquist thermal noise, which is the electrical equivalent of Planck's blackbody radiation. The noise power in Watts is given by $P = k_B T \Delta f$, where k_B is the Boltzmann constant, T the temperature and Δf the frequency bandwidth. The magnitude of the noise is shown in Fig. 4.14.

The third noise source is vibrational noise, produced mainly by rotating parts, e.g. pumps. It is in the order of a few Hz to a few hundred Hz and can be minimized by vibrational low pass filters like a heavy stone ore metal plate and by reducing the amount of connectors from the sample to the amplifier.

Fig. 4.14 Power of the
Johnson-Nyquist noise as a
function of the bandwidth at
300 K (*black*), 4 K (*red*) and
100 mK (*blue*). Notice that 0
dBm corresponds to 1 mW

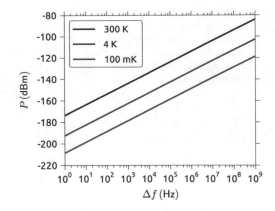

4.6.1 Low Frequency Filters

To protect the experimental setup from electromagnetic radiation, every incoming
and outgoing wire was shielded. Additionally, we tested low-pass filters, which
can be mounted at the 4 K stage to further attenuate the remaining electromagnetic
noise. They should have a negligible series resistance in order to be compatible with
the electromigration (see Sect. 4.9.2). For this reason, we were looking for suitable
pi-filters, consisting of two capacitors and one inductor. Their cut-off frequency f_0
should be around 1 MHz at cryogenic temperatures in order to have enough bandwidth for the electromigration technique. Since they will be mounted inside the
cryostat, their size should of course be as small as possible. For testing purposes,
we ordered several pi-filters with about equal size and cut-off frequency. The room
temperature transfer function s_{12} is shown in Fig. 4.15. Their cut-off frequency is
about 500 kHz with an initial attenuation of 20 dB/decade, which is increasing to
40 dB/decade at around 5 MHz. The kink in s_{12} is due to asymmetrical capacitors.
In order to test their cryogenic compatibility, we performed ten temperature cycles
from 300 to 77 K by repeatedly immersing them in liquid nitrogen. The final transfer
function s_{12} after the tenth cycle at 77 K is shown in Fig. 4.16.
All devices show a shift of f_0 to higher frequencies at 77 K. This is due to a decreasing
susceptibility ϵ_r of the dielectric with temperature. Only pi-filters from EMI Inc. with
the X7R dielectric showed an acceptable temperature stability and were therefore
selected for our setup.

4.6.2 High Frequency Filters

The attenuation of noise frequencies above 1 GHz requires a different type of filter
since discrete filters like an LC-circuit become transparent due to parasitic effects [5]
(see Figs. 4.15 and 4.16). Over the last decades, a diversity of solutions has been
proposed. The most common high frequency filters are fine-grain metal powder

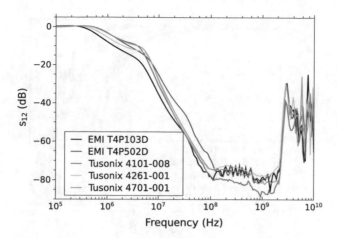

Fig. 4.15 Transfer function s_{12} at 300 K for several pi-filter measured with an Agilent E8362C vector network analyzer

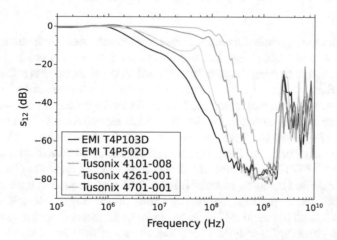

Fig. 4.16 Transfer function s_{12} at 77 K for several pi-filter measured with an Agilent E8362C vector network analyzer

filters [6–9], whose attenuation is based on skin-effect damping. Those filters are often bulky but have a very high performance. Moreover, thin coaxial cables, like mircocoax [10] or Thermocoax [11] have been tested. They are less space consuming but their attenuation is also smaller. A different approach involves lithographically fabricated meander lines, which work as distributed LRC filters [12–14]. Very recently, wires surrounded by Eccosorb, which is a microwave absorbing material, were testes under cryogenic conditions [6]. A nice summary of different filter types is given in [15].

In order to be space-efficient, we could use either the Thermocoax or Eccosorb filters. To compare the two filter techniques, we fabricated different measurement

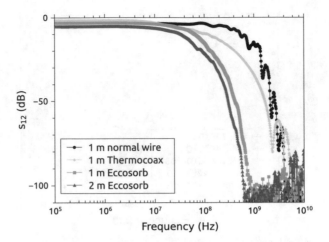

Fig. 4.17 Comparison of the attenuation of different filters. The Eccosorb filters were made out of Eccosorb-coated wires enclosed in a CuNi tube with 1.5 mm in diameter

lines, which were terminated by SMA connectors on both sides. Their attenuation was determined with the Agilent E8362C vector network analyzer. The results can be seen in Fig. 4.17. While at lower frequencies ($\approx$10 MHz) the attenuation is almost similar, the Eccosorb coated lines reach -70 dB attenuation already at around 600 MHz, the Thermocoax, however, only at around 2 GHz. As a comparison we measured also a 1 m long line without Eccosorb, which showed as expected the worst performance (see Fig. 4.17). Based on these results, we chose Eccosorb coated wires as high frequency attenuators. The final filter was made out of 24 superconducting wires made out of NbTi filaments embedded in a constantan matrix. They were coated with Eccosorb, and enclosed in a CuNi tube of 1.5 mm external diameter. The first meter of the tube is gradually thermalized from 300 K down to 40 mK, while the rest is thermalized to the 40 mK stage to attenuate all thermal noise sources. To be more space efficient, the very low temperature part of the filter was rolled up in a counterwind cylindrical coil. We chose the superconducting wires in order to keep the series resistance low, which is of paramount importance for the electromigration technique (see Sect. 4.9.2). The constantan matrix and the CuNi tube are needed to keep the heat leak from 300 K to 40 mK small. The attenuation of the final filter is shown in Fig. 4.17 (blue curve).

4.7   Signal Transducer

In Sect. 4.6, it was already pointed out that major noise sources at room temperature are electromagnetic radiation and vibrations. They couple to the experimental setup via ground loops, weak shielding or bad connectors. A way to curtail these problems is to use short cables and avoid connectors wherever it is possible. Therefore, we wanted to unify the commonly used switch box, amplifier, voltage divider, and low pass filters in one signal transducer. The development was done in close collaboration with Daniel Lepoittevin from the Néel Institute.

The signal transducer was designed to be compatible with the standard dilution fridge interface (12 pin Jaeger connector) and the batches of electromigration junctions, which have all a common source and gate, respectively (see Sect. 4.9.2). Due to the geometry of the 12 pin Jaeger connector, we ended up with 10 selectable signal injections lines (drains), one signal output line (source), and one gate. Every line can be grounded directly or via a 100 kΩ resistor. This prevents large discharge currents during the installation of the chip carrier in the cryostat, which are caused by a potential difference between the junctions and the dilution fridge. The drain and gate line have additional voltage dividers in order to increase the resolution of the data acquisition unit (see Sect. 4.8). In addition, an offset of ± 2.5 V or ± 5 V can be superimposed to the divided gate signal in order to shift the measuring range by keeping the resolution constant. To avoid sharp transitions between different offsets, a low pass filter with a time constant of 1 s is added to the summing amplifier. Furthermore, all inputs are equipped with a low pass filters to reject the incoming noise. Thereby the drain inputs have a cutoff frequency of 500 Hz and the gate input a cutoff frequency of 200 Hz. The higher value of the drain inputs was needed in order to transmit the lock-in signal, which can is modulated up to a few hundred Hz. In the following a more technical description to the signal transducer is given.

The signal transducer contains two built-in IV converters. The OPA129U (box 2 Fig. 4.18) is an ultra low input bias current amplifier. It has a current input bias of only 30 fA and is used for our actual measurements. It provides four selectable gains (R40-R43), which are $10^6 - 10^9$. Parallel to R40-R43 are the capacitors C57-C60, which on the one hand prevent the amplifier from self-oscillation and on the other hand determine its bandwidth.

To adjust the offset of the OPA129U the circuit in box 3 Fig. 4.18 is added. It consists of a very stable current source (Ref200AU), which yields a current of ± 100 μA with a precision of 0.25 % for input voltages from 2.5–40 V. In the following, this current is transformed into a voltage via the resistors R48-R50 and amplified to give an offset compensation in the range of -30 to $+30$ mV. Moreover, it should be noticed that the input of the OPA129U is directly soldered to source line in order to minimize the electro-mechanical noise.

The fast feed back loop of the electromigration requires an amplifier with a large bandwidth. Therefore, a second IV converter (LT1028CS8) is mounted inside the signal transducer (box 1 Fig. 4.18). Its internal bandwidth is 75 MHz and its current input bias is 30 nA. Due to this large value, it must be disconnected during sensitive measurements since otherwise a huge part of the signal would be lost because of the input-leak current. Its gain is fixed to 10^3, which is the optimum range for the electromigration.

The switches to select the different drain terminals have 3 positions, ground, 100 kΩ via ground, and floating. The first two positions are used when connecting the sample to the cryostat, whereas the latter is used during the experiment. The polarizing resistor of the drain voltage divider was chosen to be only a fraction of the sample

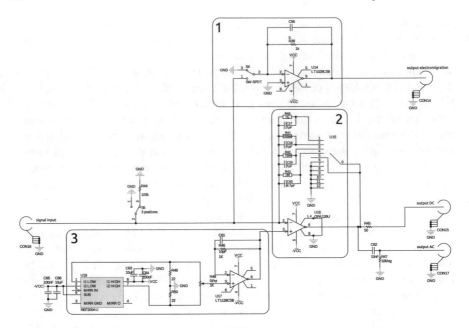

Fig. 4.18 Circuit diagram of the IV converter. Box (I) contains the fast current voltage converter with a bandwidth of 1 MHz and an amplification of 10^3. This IV converter is used for the electromigration only. Box (2) shows the ultra low noise IV converter with 4 selectable gains used for the electrical transport measurements. The offset of the IV converter in (2) is adjusted by the circuit embedded in (3)

resistance. To reject the input noise, an additional 500 Hz low pass filter was add to the drain input.

The gate-circuit divides the input voltage coming from the voltage source by up to 90 (box 3 Fig. 4.19). Afterward, an optional offset of ± 2.5 or ± 5 V is added to the divided signal using the summing amplifier (box 5 Fig. 4.19). The offset is created using the two voltage references in box 1 Fig. 4.19. The circuits in box 2 and 4 are unity gain buffer amplifiers, which transform the impedance of circuits 3 and 1 to almost zero Ω. Finally, the signal goes through a 30 Hz low pass filter (box 6 Fig. 4.19) in order to reject the low frequency noise. Figure 4.20 shows the final version of the signal transducer.

In order to analyze the performance of the high gain IV converter inside the signal transducer we were measuring its noise level with a Stanford SR760 FFT spectrum analyzer. To benchmark the results we did also measurements on the isolated IV converter and a commercial low noise IV converter Femto DLPCA-200. Therefore, we were able to verify the crosstalk to the surrounding electronics inside the signal transducer as well as the overall performance.

A scheme of the experimental setup for measuring the input and output noise is shown in Fig. 4.21a, b, respectively. Since the current noise of the inputs is very small, an additional amplifier had to be used.

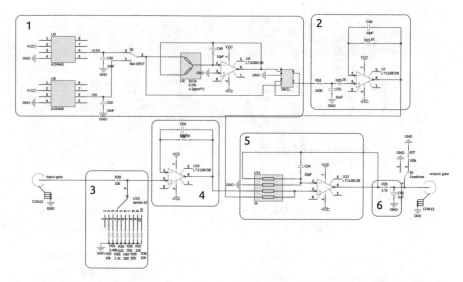

Fig. 4.19 Schematic of the gate circuit. Circuit diagram of the gate circuit. The input gate voltage is divided using the circuit in box 3. Subsequently an optional offset of 0, ±2.5 or ±5 V is generated using the circuit in box 1 and added to the divided signal using the summing amplifier in box 5. The circuit in 2 and 4 are used to match the output impedances of 3 and 1–5

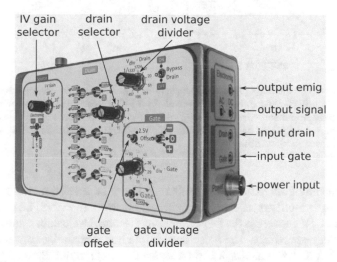

Fig. 4.20 Picture of the signal transducer

First, we disconnected the amplifier of Fig. 4.21a in order to acquire the background signal of the setup (see green curve in Fig. 4.21c). It was found to be 50 fA above the theoretical value of 128 fA ($\sqrt{4k_B T/R}$), which is most likely due to additional noise

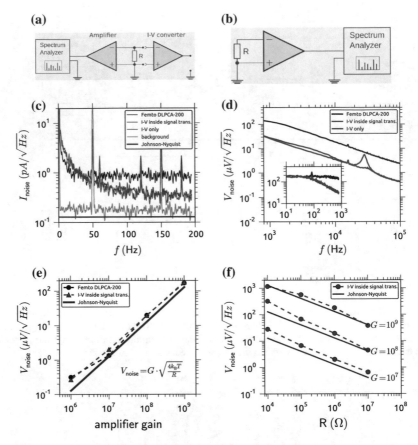

Fig. 4.21 **a** Setup for measuring the spectral input noise. The resistance R was chosen to be $1\,M\Omega$. **b** Setup for measuring the spectral output noise with $R = 1\,M\Omega$. **c** Spectral input noise measurements for three different amplifiers versus frequency f. **d** Spectral output noise of three different amplifies versus frequency f. **e** Spectral output noise versus the amplifier gain. **f** Spectral output noise versus the input resistance for three different gains

of the second amplifier. Afterward, the three different IV converters were connected to analyze their input noise. The blue, red, and black curve in Fig. 4.21 correspond to the isolated IV converter, the IV converter of the signal transducer, and the Femto DLPCA-200, respectively. The noise level of the home-made IV converter shifted to $340\,fA$ at $170\,Hz$, which is a $260\,fA$ above the background. Since the blue and the red curve are almost identical, we can exclude any crosstalk between different parts of the signal transducer with the input IV converter. If, however, the Femto DLPCA-200 is connected to the spectrum analyzer we monitor noise level of around $900\,fA$ at $170\,Hz$, which is $720\,fA$ above the background and therefore more than twice the value of the OPA129U.

In Fig. 4.21d the noise levels measured at the outputs of the amplifiers are depicted. The frequency was varied from 0 to 100+kHz, the resistor is again 1 MΩ and the gain was fixed to 10^9 V/A. At frequencies below 100 Hz, the voltage noise is almost identical for all three IV converters. The different cutoff frequencies of the two I-V converters origin from a bandwidth of 100 Hz for the OPA129U and 1 kHz for the Femto at the same gain. Hence, about one order of magnitude less high frequency noise is collected but the lock-in frequency is limited values below 100 Hz at this gain. The peak in the red curve at around 26 kHz results from a parasitic LRC oscillator inside the signal transducer. Since this peak remains below the Femto noise level its influence is considered as negligible.

In Fig. 4.21e we compared the output noise of the Femto DLPCA-200 with the IV converter of the signal transducer for different gains at a frequency of 20 Hz and with a resistor of 1 MΩ. At this frequency, the two curves are almost identical and about 1.6 times higher than the theoretical value ($S_V = G \cdot \sqrt{4k_B T / R}$).

In Fig. 4.21f the output noise of the IV signal transducer was measured for different gains and resistances. The good agreement between the obtained data and the theoretical values shows the small value of extra noise added to the signal.

4.8 Real-Time Data Acquisition

The experimental setup required the control and read-out of multiple signals simultaneously. In a straight forward realization one could use several devices, linked one to another by a common ground. This, however, induces ground loops, which would be a major source of noise. Therefore, our motivation was to combine all tasks in one automation unit like a computer. However, in conventional computers the operating system is assigning priorities to different tasks. Thus, a task with low priority can be executed with a delay of several milliseconds. Additionally, the execution of a task with high priority is not guaranteed. Hence, the simultaneous control of different experimental parameters cannot happen in a synchronized way with precisions below several milliseconds.

For this reason, we were using an ADwin system instead of a standard PC (Fig. 4.22). It combines analog and digital inputs and outputs with a dedicated real-time processor and real-time operation system. It has a 16 bit output card with an integrated D/A converter. Its voltage range is ± 10 V resulting in a step size of $20V/2^{16} = 305$ μV. The input card, in contrary, has a resolution of 18 bit and an A/D converter with readout voltages ranging from -10 and $+10$ V at a resolution of $20V/2^{18} = 75$ μV. An additional 14 bit input card with a clock frequency of 50 MHz was added to perform the electromigration using a fast feedback loop. All cards are controlled by a 300 MHz digital signal processor (DSP), which performs tasks with a precision of 3 ns. The response time in the feedback loop of the electromigration is 1.5 μs due to the execution of several lines of code.

Fig. 4.22 Picture of the ADwin automat showing the front panel with an 18 an 14 bit analog input card an a 16 bit analog output card

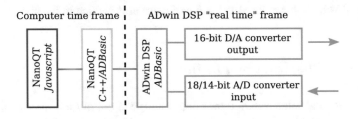

Fig. 4.23 Schematic representation of the different execution levels during the data acquisition. The user programmed Script is transcribed into different lines of C++ and ADBasic, the latter being the native language of the ADwin system. Those *lines* are send to the ADwin DSP who carries out the instructions at a frequency of 300 MHz

The ADwin is linked to a standard PC via an Ethernet connection and can be programmed using NanoQt (see Fig. 4.23). This is a home-made software, which was developed in our group by E. Bonet, C. Thirion and R. Picquerel. Its user interface is based on the JavaScript language and allows for the execution of user defined scripts.

4.9   Sample Fabrication

The device, which was studied in thesis, is a molecular spin-transistor. It consists of a single-molecule magnet, which is connected to source, drain, and gate terminals. The size of the molecule and therefore the characteristic dimension of the device was about 1 nm. Since the smallest dimensions, which can be created by electron beam lithography, are around 10 nm, other fabrication techniques were necessary.

Today, there are only a few techniques available to reliably connect a single molecule to metallic electrodes, such as a scanning tunneling microscopy [16], mechanical
break junctions [17], and electromigrated break junctions [18]. Among those techniques, electromigration is the only one which can also implement an efficient gate
to control the chemical potential of the molecule and therefore enables us to adjust
the working point of the transistor.

The first step towards a molecular spin-transistor is the fabrication of a Nanowire
with a well define weak point. Using electromigration in the next step enables us to
craft a nanometer sized gap at the predefined breaking point of the nanowire. In the
last step, we trap a molecule inside the nanogap to complete the transistor fabrication.
In the following a more detailed explanation of the three fabrication steps is given.

4.9.1 Nanowire Fabrication

The nano fabrication of our devices was done using the clean room facilities of the
Néel Institute. In order to reduce the number of external connections per transistor,
a layout with 12 nanowires sharing a common source and gate was developed. An
optical image of the layout is depicted in Fig. 4.24a. In Fig. 4.24b we can clearly see
the 12 nanowires with their source in the middle of the image and the U-shaped gate
underneath. It was already shown by [19] that back-gated single-molecule transistors
show a very good gate response and are most compatible with the electromigration
technique.

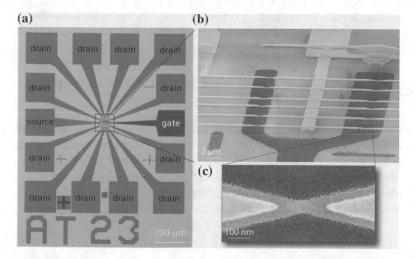

Fig. 4.24 a Layout of an array of 12 transistors sharing a common source and gate terminal. **b**
Scanning electron microscope image showing the back gate (*grey*) as well as the common source and
the different drain terminals. **c** Zoom showing the nanowire obtained by shadow mask evaporation

Fig. 4.25 Cross section of
the nanowire and the
predetermined breaking
point. The titanium sticking
layers under the gold
electrodes are not shown

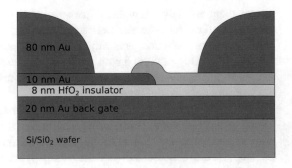

The first step in the device fabrication was the deposition of the back gate (grey
electrode in Fig. 4.24b. In consists of a 20 nm thick gold layer, which was deposited
onto a Si/SiO$_2$ wafer and a 3 nm Ti sticking layer using deep ultra violet optical
lithography and metal evaporation. During this process, the contact pads as well as
the U-shaped electrode were fabricated. To insulate the gate from the source and drain
terminals, a 8 nm thick HfO$_2$ layer with a dielectric constant of $\approx$17 was deposited
onto the gate by using atomic layer deposition. The thin oxide layer resulted also
in the different color of the gate with respect to source and drain. Subsequently, we
deposited the source and drain contact pads using ultra violet optical lithography
and Ti/Au metal deposition. The most important part for the electromigration is
the deposition of a nanowire with a predetermined breaking point. This step was
done using electron beam lithography and shadow mask evaporation under different
angles. A scanning electron microscope image of the constriction in the nanowire
is shown in Fig. 4.24c. It has thickness of only 10 nm at the weakest point, whereas
the nanowire itself is 80 nm thick. A schematic cross section of the nanowire and the
predetermined breaking point is shown in Fig. 4.25.

4.9.2   Electromigration

In order to create a nanogap between the source and drain terminal, we made use of
the electromigration technique at mK temperatures. The phenomenon of electromigration is known since a long time. Especially in the 1960s, it gained a lot of interest
since it was found to be a reason for failure of micro-electronic devices [20, 21]. The
phenomenon can be paraphrased as the diffusion of metal ions under the exposure of
large electric fields. The force applied to the each metal ion can be written as [22]:

$$\boldsymbol{F} = Z * e\boldsymbol{E},\tag{4.9.1}$$

where $Z*$ is the effective charge of the ion during the electromigration and can be decomposed into:

$$Z* = Z_{el} + Z_{wind}, \tag{4.9.2}$$

where Z_{el} can be seen as the nominal charge of the ion and Z_{wind} the momentum exchange effect between electrons and the ion, commonly also referred to as the electron wind [22]. In metals, only the latter contribution is responsible for the diffusion of the ion and not the electric field. Therefore, the diffusion happens in the direction of the electric current.

Our electromigration procedure is combination of the method of Park [18] and Strachan [23]. In order to limit the Joule heating during the electromigration, we polarize the break junction with a voltage instead of a current. The increasing resistance, which is expected during the migration of the metal, thus, leads to a power reduction (U^2/R) instead of a power increase (I^2R).

Furthermore, it was shown that a large series resistance leads to an increase of power dissipation during the electromigration [24–26], which results in larger gaps or even the complete destruction of the device. As already pointed out in Sect. 4.6.2, we were using superconducting wires inside the cryostat to reduce the total series resistance (120 Ω, measured from one connector outside the cryostat to another).

Moreover, we made us of the ADwin system to establish a fast feedback loop. It continuously reads-out the resistance of the wire and turns off the polarizing voltage within 10 μs. Since the typical time constant of the electromigration is in the order of 100 μs [27], we are able to control the size of the nanogap formation on the atomic level.

The conductance-voltage characteristic recorded during the electromigration typically looks like Fig. 4.26a. It shows a first decrease of the conductance due to Joule heating of the metal. The subsequent increase of the conductance is caused by a rearrangement of the metallic grain boundaries, which enlarge the average grain size and therefore reduced the scattering at the grain boundaries. The following sharp

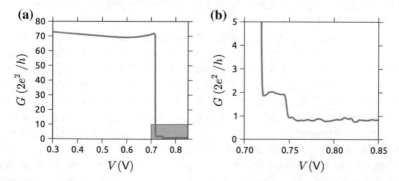

Fig. 4.26 a Conductance of the break junction during the electromigration. **b** Zoom into the *grey shaded region* of (**a**) showing quantized conductance steps

Fig. 4.27 Scanning electron
microscope image of an
electromigrated break
junction

drop in the conductance curve is caused by the migration of the gold ions, leading
to the formation of a nm sized gap. During the last seconds of the electromigration,
we are often able to see quantized conductance steps, which arise from the current
transport through the last remaining gold atoms.

A scanning electron microscope image of an electromigrated junction is presented
in Fig. 4.27. It shows the predefined breaking point of the nanowire and a nanometer
sized gap.

4.9.3 Fabrication of a Molecular Spin Transistor

Applying the procedures of Sects. 4.9.1 and 4.9.2 allows us to create a three terminal
device with source, drain, and gate electrodes. In order to complete the fabrication
of a molecular transistor, a single molecule needs to be trapped inside the nanogap,
which was formerly created by electromigration.

In the first step, we cleaned the nanowires from organic residues using acetone and
isopropanol, followed by an exposure to oxygen plasma for 2 min. Subsequently, we
dissolved 3 mg of $TbPc_2$ crystals into 5 g dichlormethane and sonicated the solution
at low power for 1 h. This ensures that the remaining $TbPc_2$ clusters are completely
dissolved. Afterward, some droplets of the solution were deposited on the nanowire
chip and blow dried with nitrogen.

In the next step, we glued the chip on the sample holder and established the electrical
connections to the chip by microbonding aluminum wires. Subsequently, the sample
was mounted inside a dilution refrigerator and cooled down to mK temperatures.
Once the sample was cold, we started the electromigration to craft a nanometer gap
into the nanowire. The heat created during this process enables the molecules to
diffuse on the surface and therefore be trapped inside the gap. As a first indication
if the fabrication procedure was successful, we measured the zero bias conductance
through the device as a function of the gate voltage. If a nanometer size object
was trapped inside the nanogap, it will create a quantum dot resulting in one or
more Coulomb peaks (see Chap. 2). Yet, this is not a proof that we actually trapped
a single $TbPc_2$ molecule. Especially when using electromigration, there are many

ways of creating a quantum dot. For example a gold nanoparticle or some organic residue, which was not completely removed during the cleaning procedure, would result in to the same transport signature when trapped inside the nanogap. In order to eliminate any doubt if the nanoparticle is a single $TbPc_2$ or not, we studied the magnetic properties of our device. As it will be shown Chap. 5, the $TbPc_2$ has a very unique magnetic signature, which can be used as a fingerprint of the molecule. In case we did not trap any or too many nanoparticles, we heated up the cryostat above 150 K and cooled it down again. This enables the surface diffusion of the molecules due to thermal activation and a subsequent retrapping at a different place. This procedure of warming up and cooling down was repeated up to ten times before changing the sample.

References

1. F. Pobell, J. Brooks, *Matter and Methods at Low Temperatures*, vol. 45, 2nd edn. (Springer, Heidelberg, 1992). ISBN 10 3-540-46356-9
2. W. Demtröder, *Experimentalphysik 1: Mechanik und Wärme (Springer-Lehrbuch) (German Edition)* (Springer, 2005). ISBN 354026034X
3. C. Boulder, V.J. Johnson, *A compendium of the properties of materials at low temperatures: (Phase 1-)*. Wright Air Development Division, Air Research and Development Command, U.S. Air Force (1961)
4. C.Y. Ho, R.W. Powell, P.E. Liley, *Thermal Conductivity of the Elements*, vol. 1. Amer Inst of Physics (1974)
5. M.M. Freund, T. Hirao, V. Hristov, S. Chegwidden, T. Matsumoto, A.E. Lange, Compact low-pass electrical filters for cryogenic detectors. Rev. Sci. Instrum. **66**, 2638 (1995)
6. S. Mandal, T. Bautze, R. Blinder, T. Meunier, L. Saminadayar, C. Bäuerle, Efficient radio frequency filters for space constrained cryogenic setups. Rev. Sci. Instrum. **82**, 024704 (2011)
7. A. Lukashenko, A.V. Ustinov, Improved powder filters for qubit measurements. Rev. Sci. Instrum. **79**, 014701 (2008)
8. F.P. Milliken, J.R. Rozen, G.A. Keefe, R.H. Koch, 50Ω characteristic impedance low-pass metal powder filters. Rev. Sci. Instrum. **78**, 024701 (2007)
9. J. Martinis, M. Devoret, J. Clarke, Experimental tests for the quantum behavior of a macroscopic degree of freedom: the phase difference across a Josephson junction. Phys. Rev. B **35**, 4682–4698 (1987)
10. D.C. Glattli, P. Jacques, A. Kumar, P. Pari, L. Saminadayar, A noise detection scheme with 10 mK noise temperature resolution for semiconductor single electron tunneling devices. J. Appl. Phys. **81**, 7350 (1997)
11. A.B. Zorin, The thermocoax cable as the microwave frequency filter for single electron circuits. Rev. Sci. Instrum. **66**, 4296 (1995)
12. D. Vion, P.F. Orfila, P. Joyez, D. Esteve, M.H. Devoret, Miniature electrical filters for single electron devices. J. Appl. Phys. **77**, 2519 (1995)
13. H. Courtois, O. Buisson, J. Chaussy, B. Pannetier, Miniature low-temperature high-frequency filters for single electronics. Rev. Sci. Instrum. **66**, 3465 (1995)
14. H. le Sueur, P. Joyez, Microfabricated electromagnetic filters for millikelvin experiments. Rev. Sci. Instrum. **77**, 115102 (2006)
15. K. Bladh, D. Gunnarsson, E. Hürfeld, S. Devi, C. Kristoffersson, B. Smalander, S. Pehrson, T. Claeson, P. Delsing, M. Taslakov, Comparison of cryogenic filters for use in single electronics experiments. Rev. Sci. Instrum. **74**, 1323 (2003)

16. J. Schwöbel, Y. Fu, J. Brede, A. Dilullo, G. Hoffmann, S. Klyatskaya, M. Ruben, R. Wiesen-danger, Real-space observation of spin-split molecular orbitals of adsorbed single-molecule magnets. Nat. Commun. **3**, 953 (2012)
17. M.A. Reed, Conductance of a molecular junction. Science **278**, 252–254 (1997)
18. H. Park, A.K.L. Lim, A.P. Alivisatos, J. Park, P.L. McEuen, Fabrication of metallic electrodes with nanometer separation by electromigration. Appl. Phys. Lett. **75**, 301 (1999)
19. H. Park, J. Park, A.K. Lim, E.H. Anderson, A.P. Alivisatos, P.L. McEuen, Nanomechanical oscillations in a single-C60 transistor. Nature **407**, 57–60 (2000)
20. I.A. Blech, Direct transmission electron microscope observation of electrotransport in aluminum thin films. Appl. Phys. Lett. **11**, 263 (1967)
21. J.R. Black, Electromigration failure modes in aluminum metallization for semiconductor devices. Proc. IEEE **57**, 1587–1594 (1969)
22. K. Tu, Electromigration in stressed thin films. Phys. Rev. B **45**, 1409–1413 (1992)
23. D.R. Strachan, D.E. Smith, D.E. Johnston, T.H. Park, M.J. Therien, D.A. Bonnell, A.T. Johnson, Controlled fabrication of nanogaps in ambient environment for molecular electronics. Appl. Phys. Lett. **86**, 043109 (2005)
24. H.S.J. van der Zant, Y. Kervennic, M. Poot, K. O'Neill, Z. de Groot, J.M. Thijssen, H.B. Heer-sche, N. Stuhr-Hansen, T. Bjørnholm, D. Vanmaekelbergh, C.A. van Walree, L.W. Jenneskens, Molecular three-terminal devices: fabrication and measurements. Faraday Discuss. **131**, 347 (2006)
25. M.L. Trouwborst, S.J. van der Molen, B.J. van Wees, The role of Joule heating in the formation of nanogaps by electromigration. J. Appl. Phys. **99**, 114316 (2006)
26. T. Taychatanapat, K.I. Bolotin, F. Kuemmeth, D.C. Ralph, Imaging electromigration during the formation of break junctions. Nano Lett. **7**, 652–656 (2007)
27. K. O'Neill, E.A. Osorio, H.S.J. van der Zant, Self-breaking in planar few-atom Au constrictions for nanometer-spaced electrodes. Appl. Phys. Lett. **90**, 133109 (2007)

Chapter 5
Single-Molecule Magnet Spin-Transistor

One of the major motivations to study single molecule magnets (SMMs) is to design ultra dense data storage devices, where each bit of information is stored on the magnetization of a single molecule. However, due to the tiny magnetic moment of an SMM (few μ_B) and a size in the order of a nanometer, it is impossible to study isolated SMMs with standard magnetometers like a micro-squid.

Therefore, we were using a completely new type of detection device—a single-molecule magnet spin-transistor. It was fabricated using electromigration of a nanowire at mK temperatures (see Sect. 4.9). In this way, a nanometer sized gap was crafted between two very clean gold terminals, in which we trapped a single TbPc$_2$ molecule magnet. An artistic view of the device is shown in Fig. 5.1. By studying the electronic transport through the device as a function of the external magnetic field, we are able to read-out the electronic spin state of an isolated single-molecule magnet and the nuclear spin state of a single terbium ion. The latter will be briefly discussed at the end of this chapter and in more detail in Chaps. 6 and 7. This chapter will mainly focus on operation of the spin-transistor and how it can be exploited to study the electronic spin of a single-molecule magnet. The terminology "spin" when it is used alone will always refer to the electronic spin.

5.1 Mode of Operation

The first working molecular spin-transistor was fabricated 2012 in our group [1] and is referred to as sample A. Later on, I fabricated two other devices, which will be referred to as sample B and C. In order to explain the working principle of the molecular spin-transistor, we will schematically subdivide the device into three quantum systems, namely, a nuclear spin qubit, an electronic spin, and a read-out quantum dot (Fig. 5.1b).

(I) The **nuclear spin qubit** emerges from the atomic core of Tb^{3+} ion. It possesses a nuclear spin of $I = 3/2$ resulting in four different qubit states. Due the hyperfine

© Springer International Publishing Switzerland 2016

S. Thiele, *Read-Out and Coherent Manipulation of an Isolated Nuclear Spin*, Springer Theses, DOI 10.1007/978-3-319-24058-9_5

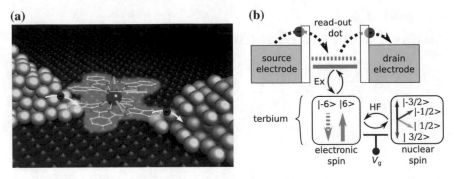

Fig. 5.1 a Artist view of the single-molecule magnet spin-transistor. The ligands of the TbPc$_2$ are tunnel coupled to the source and drain thus creating a quantum dot, which can be controlled by a back gate (not shown) underneath. In the center of the molecule is a Tb^{3+} ion possessing an electronic spin (*orange*) and a nuclear spin (*green*). **b** Simplified coupling scheme of the spin-transistor. It consists of three quantum systems: a nuclear spin qubit, an electronic spin and a read-out quantum dot. The nuclear spin is coupled with the electronic spin via the hyperfine interaction. This quantum mechanical link can be used to map the nuclear spin state onto the electronic spin, which amplifies the magnetic signal by $\approx 10^3$. Furthermore, the electronic spin is exchange coupled to the read-out quantum dot, which establishes the detection of the electronic spin and therefore nuclear spin qubit

interaction and the nuclear quadrupole moment, the degeneracy of the four levels is lifted, resulting into four unequally spaced levels (for further details see Sect. 3.7).

(II) The **electronic spin** arises from the terbium's 4f electrons. The intrinsic spin-orbit coupling and a strong ligand-field results in an electronic ground state doublet of $m_J = \pm 6$ and an easy axis of magnetization perpendicular to the ligand plane. This means that the electronic spin can be regarded as a two level system with its eigenstates $|\uparrow\rangle$ and $|\downarrow\rangle$. The degeneracy of the doublet is lifted by the hyperfine coupling to the nuclear qubit and splits each state into four levels, which are separated by approximately 2.5, 3.1 and 3.7 GHz [2]. For further details we refer to Sects. 3.5–3.7.

(III) The **read-out quantum dot** is created by the phthalocyanine (Pc) ligands. Their delocalized π-electron system is tunnel-coupled to the source and drain terminals, thus creating a conductive island. Furthermore, a finite overlap of the π-electron system with the terbium's 4f wave functions gives rise to an exchange coupling between the read-out dot and the electronic spin.

Using this device we were able to read-out the electronic spin state of the TbPc$_2$. Due to the exchange coupling between the read-out dot and the electronic spin, a slight modification in the read-out dot's chemical potential is created depending on whether the electronic spin points parallel or antiparallel to the external field. Since the position of the chemical potential with respect to the source and drain Fermi levels determines the conductance through the device, the two electronic spin states can be assigned to two different conductance values. Therefore, an electronic spin transition from $|\uparrow\rangle \rightarrow |\downarrow\rangle$ or $|\downarrow\rangle \rightarrow |\uparrow\rangle$ results in a conductance jump.

Furthermore, we can use the device to perform a single-shot read-out of the nuclear spin-qubit state. In contrary to the electronic spin detection, this is a two stage process, which takes advantage of the coupling between all three quantum systems. In the first stage, the nuclear qubit state is mapped onto the electronic spin using the hyperfine interaction. As already pointed in Chap. 3, the ligand field mixes the two electronic ground states, resulting in an anticrossing of $\Delta E \simeq 1 \, \mu K$ close to zero magnetic field. Sweeping the magnetic field slowly enough over such an anticrossing gives rise to the quantum tunneling of magnetization (QTM), which reverses the electronic spin according to the Landau-Zener probability. Due to the hyperfine interaction we get four instead of one anticrossing, which makes the magnetic field position of the QTM transition nuclear spin dependent (see Fig. 5.8a). In the second stage, we read-out the position of the QTM event through a jump in the read-out dot's conductance and establish in this way the detection of the nuclear spin-qubit state.

In the following sections, we will show step by step which experiments were conducted and what conclusions were drawn in order to derive to model explained above.

5.2 Read-Out Quantum Dot

The first experiments we performed after the electromigration of the nanowire were low-temperature electronic transport measurements. Those were used to check whether a nanometer object was trapped inside the nanogap. If so, we expected the object to behave as a quantum dot coupled to the source and drain terminals, which would result in the typical single-electron tunneling (SET) characteristics (see Chap. 2).

To check for the SET behavior we measured the conductance through the transistor as a function of the source-drain voltage V_{ds} and the gate voltage V_g. This way, a two dimensional map (stability diagram) like in Fig. 5.2a is obtained, where regions with high conductance are colored in red and the regions of low conductance are colored in blue. This stability diagram originated from device B. Figure 5.2a shows only one charge degeneracy point (CDP) within a wide gate voltage window. This is an indication of a relatively large charging energy, thus, the read-out dot must be very small. This is consistent with the claim that the quantum dot is created by the Pc ligands, but does not yet prove our model.

Furthermore, we observed a faint Kondo ridge to the left of the CDP, which indicated an odd number of electrons on the quantum dot and good coupling of the molecule to the source and drain terminals. The occurrence of a Kondo peak was observed in all three devices, indicating that a good coupling to the electrodes is probably a requirement for a functional molecular spin-transistor. The stability diagrams of the other two samples are shown in Fig. 5.2b and c for samples A and C respectively. They were measured for a smaller V_{ds} window in order to protect the devices from damage.

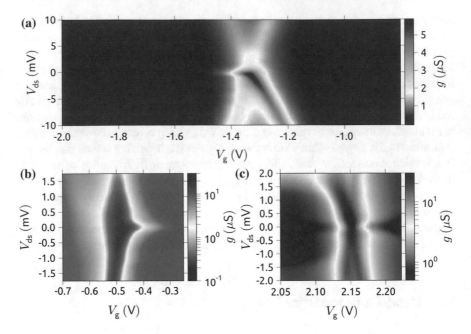

Fig. 5.2 Stability diagram of the read-out quantum dot for sample B (**a**), sample A (**b**) and sample C (**c**). They were measured by sweeping source-drain voltage V_{ds} at different gate voltages V_g while monitoring the conductance. They all show single-electron tunneling and a large Coulomb blockade effect, which was expected from electronic transport through a single molecule, tunnel-coupled to source and drain electrodes. Furthermore, a Kondo peak was observed for all devices, indicating a good coupling to the source and drain terminals. Note that the exotic appearance of the Kondo peak in (**c**) will be discussed in more detail in Sect. 5.4

5.3 Magneto-Conductance and Anisotropy

A first test to verify if the quantum dot, presented in the previous section, was coupled to the magnetic moment of the TbPc$_2$ molecule, is to study the conductance through the device as a function of the magnetic field. Since the magnetic moment of the terbium double-decker can be reversed with an external magnetic field, we expected to see a feature of this magnetization reversal in the electronic transport. To perform the magneto-conductance measurement, we fixed V_{ds} at zero Volt and V_g at a value close to the charge degeneracy point. That is where we expected the largest sensitivity to a magnetization reversal, as a slight variation of the quantum dot's chemical potential results in a strong modification of the conductance. Afterward, we swept the external magnetic field from negative to positive values (trace) and back again (retrace) while recording the conductance through the quantum dot.

As shown in Fig. 5.3a, by sweeping the magnetic field back and forth, we observed jumps in the read-out dot's conductance. Moreover, the magneto-conductance signal was hysteretic, which is the signature of an anisotropic magnetic object. Every

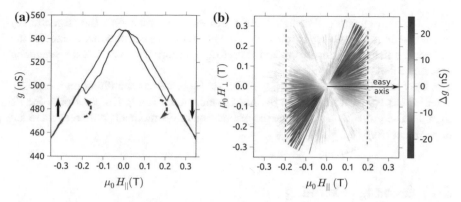

Fig. 5.3 **a** Trace (*blue*) and retrace (*red*) magneto-conductance signal of sample B as a function of $H_{||}$. The conductance jumps correspond to the reversal of the electronic spin carried by the TbPc$_2$ SMM. **b** Two dimensional magneto-conductance signal of sample B as a function of the external field recorded with an angular resolution of 0.5°. The magnetization reversal is shown as a sharp color change. The applied field to reverse the magnetization is smallest along $H_{||}$ and augments gradually with increasing angle. At an angle of 90° the magnetic field is applied in the hard plane and the magnetic moment cannot be reversed anymore

time this object reversed its magnetization the chemical potential of the read-out dot changed between two distinct values, giving rise to jumps in the conductance. The amplitude of the jump was about 3 % of the total conductance value and approximately the same for all three devices.

In order to find more proofs that those conductance jumps originated from the spin reversal of the TbPc$_2$ SMM, we investigated the angular dependence of those jumps. In Chap. 3 we pointed out that an isolated TbPc$_2$ molecule possesses a strong magnetic anisotropy, with an easy axis of magnetization perpendicular to the phthalocyanine plane. It was shown by spin resolved DFT calculations that this anisotropy is preserved even when the molecule is brought to contact with a metallic surface [3] and should therefore also be conserved in our spin-transistor configuration.

As it was shown in Fig. 5.3a the conductance through the read-out dot depends on the orientation of the electronic spin. While at negative $H_{||}$ the spin ground state is $| \uparrow \rangle$ and the excited state is $| \downarrow \rangle$, the Zeeman effect will inverse the energies of the two states at positive magnetic field. Therefore the observed conductance jump during the trace sweep at $B \approx 0.2$ T in Fig. 5.3a corresponds to the transition $| \uparrow \rangle \rightarrow | \downarrow \rangle$ and the jump during the retrace sweep at $B \approx -0.2$ T to the transition $| \downarrow \rangle \rightarrow | \uparrow \rangle$. We repeated the hysteresis measurement under different angles of the magnetic field, and thus scanned the magneto-conductance signal within a plane in the three dimensional vector space. Between two subsequent sweeps, the vector of the magnetic field was rotated by 0.5°. Notice that the specific orientation of the plane was chosen prior to the experiment in order to include the easy axis of magnetization. Subtracting the retrace signal from the trace signal at each angle resulted in Fig. 5.3b. The sharp color change from white to red/blue indicates the spin reversal. By looking at the angular dependence of the reversal it is evident that it becomes harder to flip the spin, as we turn the magnetic field from $H_{||}$ towards $H_\perp$ since only the projection of the magnetic field onto $H_{||}$ is relevant. This behavior is a direct consequence of the molecule's magnetic anisotropy and therefore a strong evidence that the magnetic object is a TbPc$_2$ SMM.

The easy axis of magnetization is parallel to $H_{||}$, whereas the direction along $H_\perp$ is called hard axis. It lies within the hard plane, which is aligned parallel to the Pc ligands. Therefore, we can deduce the orientation of the molecule with respect to the experiment from Fig. 5.3b.

5.4 Exchange Coupling

In the previous section we stated that the origin of the magneto-conductance signal was due to a coupling between the read-out quantum dot and the molecule's electronic spin. In this section we are going to determine the strength of the coupling and discuss the possible origins.

To estimate the magnitude of the coupling we investigated the evolution of the Kondo peak of Fig. 5.2(a) (sample B) as a function of the bias voltage V_{ds} and the applied magnetic field B_z. Figure 5.4(a) shows that the Kondo peak is splitting linearly at a rate of 223 μV/T with augmenting B, which is expected for a spin 1/2 and a g-factor close to two.

By extrapolating the linear slopes at positive magnetic fields to negative fields, we found an intersection at approximately -210 mT, which is equal to a negative critical field B_c (see Sect. 2.4). This is in contrast to the classical spin 1/2 Kondo effect, where

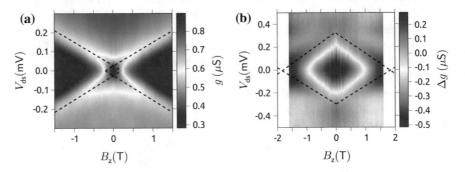

Fig. 5.4 a The conductance of sample B, at the *left side* of the charge degeneracy point, is measured as a function of the source-drain voltage V_{ds} and the external magnetic field B_z. It shows the linear evolution of the Kondo peak with respect to the magnetic field. The extrapolation of the linear slops shows an intersection at ± 200 mT, which can be used to estimate the magnitude of the coupling between the read-out dot and the electronic spin. **b** Conductance measurement of sample C showing a linear decrease of the Kondo splitting with increasing magnetic field amplitude B_z. This feature is a signature of an antiferromagnetic coupling of the terbium's electronic spin to the read-out quantum dot

the B_c is always positive and linked to the Kondo temperature T_K via: $2g\mu_B B_c = k_B T_K$. In order to explain this finding, we used the analog to the underscreened spin 1 Kondo effect (see Sect. 2.4), where the antiferromagnetic coupling between the screened spin 1/2 and the electrons in the terminals is weakened by a ferromagnetic coupling to the unscreened spin 1/2, which decreases the critical field from finite values to almost zero Tesla [4]. In our device the negative B_c can be interpreted as a ferromagnetic coupling between the read-out dot and the terbium's electronic spin. Due to the larger magnetic moment of $9\mu_B$, the antiferromagnetic coupling to the leads is already destroyed at zero bias. To model the magnetic field behavior, we modified the above mentioned formula to [1]:

$$2g\mu_B B_c = k_B T_K + a\, g\mu_B J_z \qquad (5.4.1)$$

where a is a negative for ferromagnetic coupling. From Fig. 5.4a we obtained the full width at half maximum of the Kondo peak at $B = 0$ T of 56 µV. Using the expression $eV = k_B T_K$, we get an estimated Kondo temperature to 650 mK. By inserting the Kondo temperature and $B_c = -210$ mT into Eq. 5.4.1 we extracted a coupling constant of $a = -200$ mT, indicating a strong ferromagnetic coupling.

The same experiments were performed on sample C (Fig. 5.4b) and sample A. Also in these two samples a splitting of the Kondo peak at zero magnetic field was observed. In contrary to sample A and B, sample C shows an antiferromagnetic coupling to the quantum dot since the splitting decreases for increasing B_z (see Fig. 5.4b). This behavior was modeled using a positive a in Eq. 5.4.1. Moreover, it demonstrated that the sign of the coupling constant a must be very sensitive to local deformations of the molecule, which are different from sample to sample. The coupling strengths of samples A, and C were extracted following the same procedure as explained above. Table 5.1 summarizes the three different values.

The modulus of the coupling is very large in all three samples, which makes the exchange interaction the most likely candidate. It has been demonstrated that it can attain values up to 7 T for a nitrogen atom inside a C_{60} [5] and it was presented that the TbPc$_2$ molecule shows an antiferromagnetic super-exchange coupling when deposited on a ferromagnetic surface [6].

Moreover, by considering the magnetic moment of the terbium ($9\mu_B$) and the average distance between the terbium ion and the electron on the phthalocyanine (0.7 nm), we can estimate the dipole-dipole interaction to be about 50 mT, which is smaller than the measured interaction and not sufficient to explain the coupling strength.

Table 5.1 Summary of the extracted coupling strengths of the electronic spin to the read-out quantum dot

Sample	A	B	C
a	-300 mT	-200 mT	$+1.66$ T

Yet, the exchange coupling is only possible if the read-out quantum dot and the terbium ion are geometrically very close to each other. This supports the assumption that the read-out quantum dot is created by the phthalocyanine ligands. Note that adding one electron to the read-out dot will not affect the charge state of the Tb ion, since this would require an oxidation or reduction of the terbium. It was shown by Zhu et al. [7] that up to the fifth reduction and second oxidation of the molecule, electrons are only added to the organic ligands of the double-decker, leaving the charge state and therefore the magnetic properties of the terbium ion untouched.

5.5   2D Magneto-Conductance of the Read-Out Dot

After having quantified the strength of the coupling between the read-out quantum dot and the electronic spin, we now investigate the magneto-conductance signal along two different directions. Using sample A, we measured the conductance through the read-out dot as a function of the magnetic field $H_\perp$ perpendicular to the easy axis of the TbPc$_2$ at four different parallel fields $H_\parallel$ and vice versa (Fig. 5.5).

In order to assign the electronic spin state to a certain conductance value, we fitted the data to the empirical formula

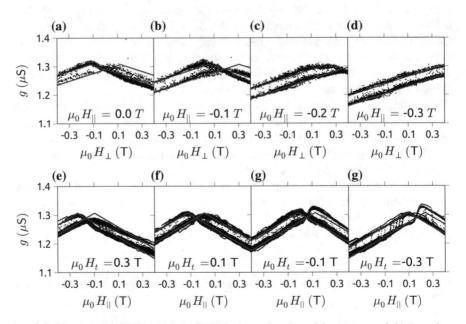

Fig. 5.5 Magneto-conductance signal of sample A as a function of the transverse field $B_\perp$ at four different parallel fields $B_\parallel$ (**a–d**) and vice versa (**e–f**). The *red curve* is an empirical fit to the spin-down conductance and the blue curve to the spin-up conductance

$$g(B_\parallel, B_t) = -\alpha|B_t - \beta B_\parallel \pm \gamma/2| + g_0 \qquad (5.5.1)$$

with $\alpha = 1.38 \times 10^{-7}$ S/T, $\beta = -1.8$, $\gamma = 0.25$ T and $g_0 = 1.307 \times 10^{-6}$ S. The fit to the spin-up conductance is depicted in blue, whereas the fit to the spin-down conductance was colored in red in Fig. 5.5. We observed that the difference between the two conductance values is constant over a large range in magnetic field but goes to zero at a particular combination of $H_\parallel$ and $H_\perp$. To get a better visualization of this effect we simulated the two dimensional conductance map using Eq. 5.5.1 and the parameters extracted from the fits (see Fig. 5.6a). In the red area, the conductance was larger when the spin pointed up, whereas in the blue area, the conductance was larger when the spin pointed down. A remarkable feature is, however, the white stripe, indicating that the two conductance values were equal.

To get a deeper understanding of the origin of the magneto-conductance signal we used a semi-classical model to describe the read-out dot's chemical potential. We assumed that the read-out quantum dot possesses a spin S, which is exchange coupled to the electronic spin J through aSJ. The Hamiltonian of the read-out dot exposed to an external magnetic field B_{ext} is given by:

$$H = g\mu_B S B_{ext} + aSJ = g\mu_B S\left(B_{ext} + \frac{aJ}{g\mu_B}\right) \qquad (5.5.2)$$

with g the g-factor of the quantum dot and μ_B the Bohr magneton. In the semi-classical approach J is no longer an operator but a vector with $J_z = \pm 6\hbar$ and $max(J_x) = max(J_y) = \sqrt{7}\hbar$. A magnetization reversal of the electronic spin was modeled by changing $J_x \rightarrow -J_x, J_y \rightarrow -J_y, J_z \rightarrow -J_z$. Like in the experiment the external magnetic field was simulated to be in the y-z plane with $B = (0, B_t, B_\parallel)$.

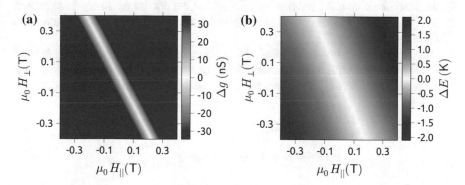

Fig. 5.6 a Fitted difference between the spin-down and spin-up conductance as function of the magnetic field parallel (∥) and transverse (⊥) to the easy axis of the TbPc₂. **b** Zeeman energy difference of the read-out quantum dot with respect to the electronic spin up or down state, calculated using Eq. 5.5.3 and $a = 200$ mT. The qualitative agreement of the two plots shows that the magneto-conductance signal can be explained by a change of the read-out dot's Zeeman splitting ΔE_Z with respect to the electronic spin state of the terbium ion

The Hamiltonian of the quantum dot is now written as:

$$H = g\mu_B \left[S_y \left(B_\perp \pm \frac{aJ_y}{g\mu_B} \right) + S_z \left(B_{||} \pm \frac{aJ_z}{g\mu_B} \right) \right]$$ (5.5.3)

with S_y and S_z being the appropriate spin matrices for the spin S and the $\pm$ sign indicating the two different spin directions $| \uparrow \rangle$ and $| \downarrow \rangle$. Assuming $S = 1/2$ we can diagonalize the Hamiltonian for each electronic spin direction individually and calculate the difference of the Zeeman splittings $\Delta E_Z(| \uparrow \rangle) - \Delta E_Z(| \downarrow \rangle)$. Doing this at different $B_{||}$ and $B_\perp$ resulted in Fig. 5.6b. The two plots in Fig. 5.6 show a good qualitative agreement, especially the white region of zero sensitivity as well as the angle with respect to $B_{||}$ is very well reproduced. It shows that the origin of the magneto-conductance signal can be explained by a shift of the quantum dot's Zeeman splitting depending on whether the electronic spin points parallel or antiparallel to the external field.

5.6 Electronic Spin Relaxation

After having explained the coupling of the read-out quantum dot to the electronic spin, we now focus on the electronic spin only. In this section we investigate the relaxation behavior of the electronic spin at large magnetic fields.

From Fig. 5.3 we can already see that the spin relaxation at large magnetic fields is not exactly determined by the projection on $H_{||}$, which origins from the stochastic nature of the inelastic spin reversal. It requires an energy exchange with the thermal bath and the creation of a phonon.

In the case of an isolated terbium double-decker the energy exchange is mediated by the ligand field. In order to quantify this effect, we will use a model taken from Abragam and Bleaney [8].

Therein, we assume a two-level spin-system whose energies are separated by $\hbar\omega$ and which is in contact with a phonon bath of temperature T. Then, the transition rates between state $|1\rangle$ and $|2\rangle$ are given by the Einstein coefficients of absorption and emission:

$$w_{1\to2} = B\rho_{ph},$$ (5.6.1)

$$w_{2\to1} = A + B\rho_{ph} = B\rho_{ph}exp\left(\frac{\hbar\omega}{k_B T} \right)$$ (5.6.2)

where ρ_{ph} is the phonon density, B the coefficient of stimulated emission or absorption and A the coefficient of spontaneous emission. If the spin-system is out of thermal equilibrium, it will return to it in a characteristic time τ:

$$\frac{1}{\tau} = w_{1\to2} + w_{2\to1}$$ (5.6.3)

which under substitution of Eqs. 5.6.1 and 5.6.2 results in:

$$\frac{1}{\tau} = B\rho_{ph}\left[exp\left(\frac{\hbar\omega}{k_BT}\right) + 1\right] \tag{5.6.4}$$

The phonon density of a three dimensional crystal is given as:

$$\rho_{ph} = \underbrace{\frac{3}{2\pi^2}\frac{\omega^2}{v^2}}_{\text{density of states}}\underbrace{\frac{\hbar\omega}{exp\left(\frac{\hbar\omega}{k_BT}\right) - 1}}_{\text{average phonon energy}} \tag{5.6.5}$$

Hence, inserting this expression into Eq. 5.6.4 gives:

$$\frac{1}{\tau} = \frac{3\hbar\omega^3}{2\pi^2v^3}B\,coth\left(\frac{\hbar\omega}{2k_BT}\right) \tag{5.6.6}$$

The lattice vibrations couple not directly to the terbium ion, instead they modulate the ligand field. To take this indirect interaction into account we develop the ligand field in powers of strain [9]:

$$V = V^{(0)} + \epsilon V^{(1)} + \epsilon^2 V^{(2)} + ... \tag{5.6.7}$$

where the first term on the right is just a static term and the second and third term correspond to first and second order corrections respectively. Applying Fermi's golden rule:

$$w_{i\to j} = \frac{2\pi}{\hbar^2}\left|\langle i| H^{(1)} |j\rangle\right|^2 f(\omega) \tag{5.6.8}$$

where $H^{(1)}$ is the first order perturbation Hamiltonian and $f(\omega)$ the normalized line-shape function. Inserting $2\rho v^2\epsilon^2 = \rho_{ph}d\omega$ and integrating over all frequencies results in $w_{i\to j} = \frac{2\pi}{\hbar^2}\frac{\rho_{ph}}{2\rho v^2}$. When we compare this expression with Eq. 5.6.1, we get $B = \frac{\pi}{2\pi\hbar\rho v^2}\left|V^{(1)}\right|^2$, with ρ being the density of the material. Hence, using Eq. 5.6.6 we get:

$$\frac{1}{\tau} = \frac{3}{2\pi\hbar\rho v^5}\left|V^{(1)}\right|^2\omega^3\,coth\left(\frac{\hbar\omega}{2k_BT}\right) \tag{5.6.9}$$

In the case of TbPc$_2$, the energy difference between the two spin states is $\hbar\omega = g\mu_B\Delta m_j\mu_0 H_{||}$. Furthermore, if $\hbar\omega \gg 2k_BT$ the hyperbolic cotangent is close to unity and the characteristic relaxation times is proportional to $(\mu_0 H_{||})^3$:

$$\frac{1}{\tau} \propto \alpha(\mu_0 H_{||})^3 \tag{5.6.10}$$

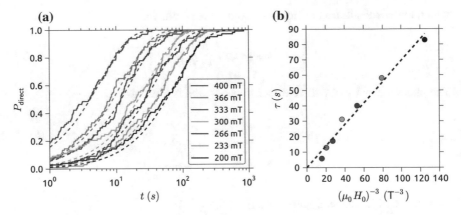

Fig. 5.7 **a** The solid lines represent the relaxation probability P_{direct} as a function of the waiting time t at different magnetic fields H_0, which are indicated in the legend. The dashed lines are a fit to $P_{direct} = 1 - exp(t/\tau)$. **b** Characteristic relaxation time τ extracted from the fits of (a) as a function of H_0

In order to verify if this model is correct within the limit of an isolated molecule, we performed the following experiment. We prepared the spin in its ground state by applying a large negative magnetic field of $\mu_0 H_{||} = -600$ mT. Afterward, we initialized the spin in its excited state by sweeping the magnetic field at 50 mT/s to $+\mu_0 H_0$, which was ranging from 200 to 400 mT. If a magnetization reversal occurred before reaching $+\mu_0 H_0$, the initialization was repeated. If, however, the spin was properly initialized in its excited stated, we recorded the time necessary to relax back into its ground state. We repeated this procedure 100 times at each H_0 and plotted the waiting times in a normalized histogram. Integrating the latter led to the extraction of the relaxation probability P_{direct} as a function of the waiting time t (see Fig. 5.7a). Subsequently each curve was fitted to the function $P_{direct} = 1 - exp(t/\tau)$ in order to obtain the characteristic relaxation time τ at each H_0. By plotting every τ as a function of $(\mu_0 H_0)^{-3}$, a straight line can be fit to the data.

This experiment is another evidence that the observed conductance jumps were indeed due to the relaxation of the electronic spin. Furthermore, the single electronic-spin quantum-system is coupled to the ligand field, which makes it behave as a classical two level system.

5.7 Quantum Tunneling of Magnetization

In the previous section, the relaxation of the magnetic moment of a single TbPc$_2$ due to a direct transition was discussed. However, the quantum nature of single molecule magnets allows for a second type of spin reversal, which is called quantum tunneling of magnetization (QTM). It was first discovered by Friedman and Thomas in 1996

[10, 11] as they measured the hysteresis loop of a Mn_{12} SMM. Henceforward, it has been extensively studied by different groups on clusters or arrays of single molecule magnets [12, 13]. Nevertheless, measuring the phenomenon on an single molecule level is quite exclusive and was first presented in 2013 using a $TbPc_2$ spin valve coupled to a carbon nanotube [14].

Before explaining the experiment, we want to recall the Zeeman diagram of the $TbPc_2$ electronic ground state doublet (see Fig. 5.8a). It shows that each electronic state was split into four levels due to the hyperfine coupling. All lines with the same slope correspond to the same electronic spin state and all lines with the same color correspond to the same nuclear qubit state. Our main focus is directed on the avoided level crossings, highlighted by colored rectangles. They were induced due to off-diagonal terms in the ligand field Hamiltonian and mix the electronic spin-up $|\uparrow\rangle$ and spin-down $|\downarrow\rangle$ state (see Sect. 3.6). However, the horizontal separation of the anticrossings is determined by the hyperfine coupling between the terbium's electronic and nuclear spin.

By applying an external magnetic field parallel to the easy axis of the molecule, we move along the lines of the Zeeman diagram. Every time we pass by one of those anticrossings, the molecule's electronic spin is able to reverse due to a process which is referred to as the quantum tunneling of magnetization (QTM). The probability of the reversal P_{LZ} is given by the Landau-Zener (LZ) formula [15, 16]:

$$P_{LZ} = 1 - exp\left[-\frac{\pi\Delta^2}{2\hbar g_J \Delta m_J \mu_0 dH_{||}/dt}\right]. \qquad (5.7.1)$$

Since this process is only allowed in the close vicinity of the anticrossing, the electronic spin can tunnel only at four distinct magnetic fields (see Fig. 5.8). Therefore, the detection of the four QTM transitions would be the final evidence that the magnetic object, which is coupled to the quantum dot, is without a doubt a single $TbPc_2$ SMM.

In order to find experimental evidence of this process, we biased the spin-transistor at $V_{ds} = 0$ and set V_g to a value in the vicinity of the charge degeneracy point. Afterward, the external magnetic field was swept from -60 to 60 mT and back while measuring the conductance through the quantum dot. The recorded magneto-conductance signal of four selected sweeps is depicted in Fig. 5.8b. It shows conductance jumps at four different magnetic fields with an amplitude of 3 % of the total conductance. The change from one conductance value to another originated from the electronic spin reversal. To demonstrate that these reversals were caused by a QTM transition, we recorded the magneto-conductance signal for several thousand sweeps. For each electronic spin reversal, we determined the magnetic field of the resulting conductance jump. By plotting the positions of all detected jumps in a histogram we obtained Fig. 5.8c and d for samples A and C respectively. More details on the data analysis are given in Sect. 6.1. We observed four nonoverlapping peaks, whose maxima coincide with the magnetic field of the four anticrossings, which is a direct evidence that the magnetic object coupled to the read-out quantum dot is a single terbium double-decker SMM. Moreover, the experiment establishes the electronic detection of the

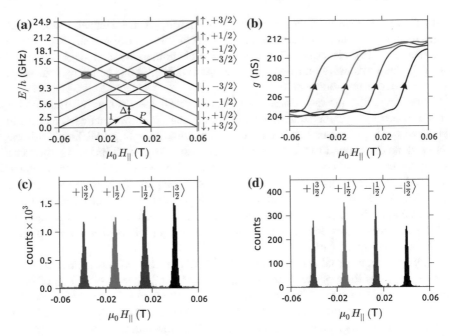

Fig. 5.8 **a** Zeeman diagram of the TbPc$_2$ molecular magnet, focusing on the isolated electronic spin ground state doublet $m_J = \pm 6$, as a function of the external magnetic field $H_{\parallel}$ parallel to the easy-axis of magnetization. Both electronic spin states $| \uparrow \rangle$ and $| \downarrow \rangle$ are split into four energy levels due to a strong hyperfine interaction with the Tb nuclear spin. The ligand field induces off-diagonal terms in the spin Hamiltonian leading to avoided level crossings (colored rectangles and inset), where quantum tunneling of magnetization (QTM) is allowed. Note that for each QTM event the nuclear spin is preserved. Therefore, the positions in magnetic field $H_{\parallel}$, where the electronic spin reversal happens, yields the nuclear spin states $| -3/2 \rangle$, $| -1/2 \rangle$, $| +1/2 \rangle$ or $| +3/2 \rangle$. **b** Magneto-conductance measurement of the read-out quantum dot. The electronic spin reversal results in a conductance jump of about 3% of the signal. **c** Histogram of all recorded conductance jumps measures on sample C. Is shows four nonoverlapping peaks originating from the QTM transitions at the avoided level crossings. They are used as a fingerprint to identify the TbPc$_2$ single molecule magnet and establish the read-out of a single nuclear spin since they link the magnetic field of the conductance jump to each nuclear qubit state. **d** Histogram similar to (**c**) measured on sample A

nuclear spin qubit since the position of each conductance jump becomes nuclear spin dependent.

In the following we present the tunnel probability P_{QTM} as a function of the sweep rate $dH_{\parallel}/dt$ using sample A and C. Focusing on the QTM probability averaged by the four anticrossings, we swept the magnetic field back and forth from -60 to $+60$ mT. Moreover, by limiting the magnetic field amplitude to 60 mT we could also suppress direct transitions whose characteristic time was extrapolated to 53 min at this field using Fig. 5.7b. For each measurement, we checked for a conductance jump indicating the QTM of the spin. By repeating this protocol 100 and 1000 times for each sweep rate and counting the amount of the detected QTM transitions, we were

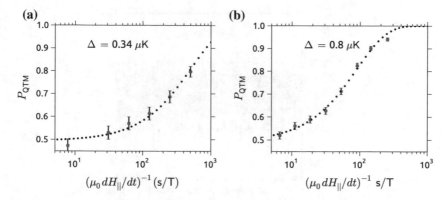

Fig. 5.9 Probability of observing a quantum tunneling of magnetization P_{QTM} of a single spin as a function of the magnetic field sweep rate $\mu_0 dH/dt$ for sample A (**a**) and C (**b**). The experimental results (*red dots*) were fitted to the function $P_{QTM} = 1 - A\,exp(B/dH_{\|}/dt)$

able to extract the tunnel probability P_{QTM} as function of $dH_{\|}/dt$ for samples A and C, respectively (see Fig. 5.9).

The results show an exponential increase of the tunnel probability with decreasing sweep rate. Fitting the data to the function $P = 1 - A\,exp(B/dH_{\|}/dt)$ enabled us to extract a tunnel splitting of $\Delta = 0.34\,\mu K$ for sample A and $\Delta = 0.8\,\mu K$ for sample C. Both values are close to the value of $1\,\mu K$ determined by Ishikawa et al. [2]. However, there is a striking deviation from Eq. 5.7.1, the tunnel probability P_{QTM} appears to converge to 50 % at high sweep rates for both samples. This implies that there must be a second process, different from the QTM, causing a reversal.

In order to learn more about the additional transition, we determined the correlation between subsequent measurements. Since the tunnel process is a random event, its correlation will vanish leaving only the additional transition for the analysis. To calculate the autocorrelation function C_n we applied the following algorithm. A spin reversal in measurement i was saved as $x_i = 1$ and no spin reversal resulted in $x_i = -1$. Subsequently the autocorrelation function was determined as

$$C_n = \frac{\sum_{i=0}^{N-n}(x_i - \bar{x})(x_{i+n} - \bar{x})}{\sqrt{\sum_{i=0}^{N-n}(x_i - \bar{x})^2}\sqrt{\sum_{i=n}^{N}(x_i - \bar{x})^2}} \tag{5.7.2}$$

with N being the total number of measurements and $\bar{x}$ mean value. Figure 5.10 shows the result of this calculation up to $n = 1000$. In order to be truly random, the correlation function must be below the $2/\sqrt{N}$ limit (red-dotted line), which corresponds to the 95 % fidelity of a random event. Since this is true, apart from some exceptions, we concluded that the additional reversal process is a random event as well and has no magnetic origin.

In the following we studied the number of transitions as a function of the source-drain offset voltage in order to analyze if additional spin reversals might be activated by the tunnel current. Therefore, we swept the magnetic field back and forth between -60

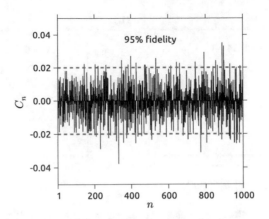

Fig. 5.10 The autocorrelation function C_n between QTM events, which were separated by n measurement (*black line*) and the 95 % fidelity threshold indicating the randomness of the autocorrelation function (*red dotted line*). Since C_n is most of the time below the threshold we could reason that QTM transitions happened at random, as expected from a quantum tunneling process

to 60 mT at 50 mT/s while gradually increasing the source-drain voltage. Every spin reversal recorded during this measurement is marked as a black point in Fig. 5.11a. It illustrates that the four peaks, corresponding to the four different nuclear spin states, were broadened with increasing bias voltage, and the appearance of additional noise in between the peaks was observed. Increasing the offset above 350 μV led to total loss of the signal. Dividing Fig. 5.11a into ten intervals of 40 μV (corresponding to 800 measurements), and integrating the number of reversals in each interval, yielded Fig. 5.11b. It shows a continuous increase of spin reversals with augmenting bias, which demonstrates the activation of the spin reversal due to the tunnel current. We suppose that the mechanism is similar to the one presented by Heinrich et. al [17], where tunnel electrons having an energy larger than the Zeeman splitting of a manganese atom are able to flip its spin. The difference in our case, is that the 4f electrons of the Tb are not directly exposed to the tunnel current as it is the case of the 3d electrons in the manganese. Therefore, we belief that the effect is less pronounced and thus less efficient.

To find out more about the activation probability, we subtracted a histogram of 1000 measurements at 300 μV offset from a histogram acquired at zero offset (see Fig. 5.11c). The result still exhibits four peaks, albeit broadened, which suggests that the activated spin reversal had a higher probability in the vicinity of the four avoided level crossings and therefore at smaller energy gaps.

Using this results, we can explain the convergence of the QTM probability to 50 %. In order to obtain a decent signal to noise ratio, lock-in amplitudes of 250 μV were necessary. From Fig. 5.11a we see that those amplitudes are already sufficient to activate the spin reversal around the avoided level crossing. This, in turn, would lead to additional transitions at each avoided level crossing, which resemble QTM events. A definite answer, however, requires theoretical modeling and is left as a future project.

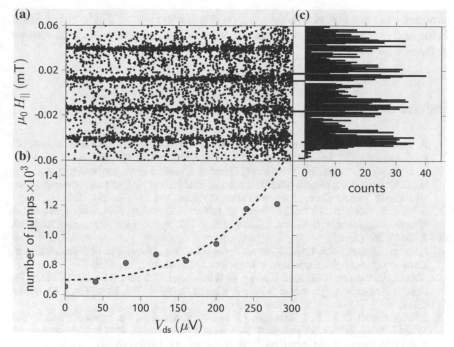

Fig. 5.11 **a** Electronic spin reversal as a function of the magnetic field $H_{||}$ and the offset bias voltage V_{ds}. **b** Integrated number of spin reversals of (**a**) for intervals of 800 sweeps. The steady increase of reversals with augmenting V_{ds} demonstrated the an activation of the reversal due to the tunnel current and is assumed to be the reason of the offset of the QTM probability. **c** Difference between a histogram of 1000 measurements taken at 300 μV offset and zero offset. The perceptibility of the four peaks shows that the activation probability is inverse proportional to the level splitting between up and down and therefore larges in the vicinity of the anticrossing

5.8 Summary

In this chapter we were able to show that an single TbPc$_2$ molecular magnet was trapped in between two gold contacts, allowing for the electronic read-out of the molecule's spin via a quantum dot. A close investigation of the coupling between the spin and the read-out quantum dot suggested that the latter was created by the organic ligands of the molecule. We presented a schematic model, which was able to describe the mode of operation of a single-molecule magnet spin-transistor, i.e., the read-out of the electronic spin. Furthermore, we presented a study of the electronic spin relaxation at large magnetic fields ($B > 200$ mT) and could extract a field dependent relaxation time $\tau(B_{||}) = 0.7(\mu_0 H_{||})^{-3}$ T^{3}s. In the end, we investigated the quantum tunneling of magnetization of the electronic spin. From the experiments we extracted a tunnel splitting of 0.34 and 0.8 μK for samples A and C, which was in the same order of magnitude as the theoretical values given by Ishikawa et al. [2]. Moreover, we were able to identify the four individual QTM transitions, which is the strongest evidence that nano-object under investigation was as single TbPc$_2$ SMM. In the next chapter we will use those transitions to perform a time-resolved read-out of the nuclear qubit state.

References

1. R. Vincent, S. Klyatskaya, M. Ruben, W. Wernsdorfer, F. Balestro, Electronic read-out of a single nuclear spin using a molecular spin transistor. Nature **488**, 357–360 (2012)
2. N. Ishikawa, M. Sugita, W. Wernsdorfer, Quantum tunneling of magnetization in lanthanide single-molecule magnets: bis (phthalocyaninato) terbium and bis (phthalocyaninato) dysprosium anions. Angew. Chem. (International ed. in English) **44**, 2931–2935 (2005)
3. L. Vitali, S. Fabris, A.M. Conte, S. Brink, M. Ruben, S. Baroni, K. Kern, Electronic structure of surface-supported bis (phthalocyaninato) terbium (III) single molecular magnets. Nano Lett. **8**, 3364–3368 (2008)
4. N. Roch, S. Florens, T.A. Costi, W. Wernsdorfer, F. Balestro, Observation of the underscreened kondo effect in a molecular transistor. Phys. Rev. Lett. **103**, 197202 (2009)
5. N. Roch, R. Vincent, F. Elste, W. Harneit, W. Wernsdorfer, C. Timm, F. Balestro, Cotunneling through a magnetic single-molecule transistor based on N@C_{60}. Phys. Rev. B **83**, 081407 (2011)
6. A. Lodi Rizzini, C. Krull, T. Balashov, J.J. Kavich, A. Mugarza, P.S. Miedema, P.K. Thakur, V. Sessi, S. Klyatskaya, M. Ruben, S. Stepanow, P. Gambardella, Coupling single molecule magnets to ferromagnetic substrates. Phys. Rev. Lett. **107**, 177205 (2011)
7. P. Zhu, F. Lu, N. Pan, D.P. Arnold, S. Zhang, J. Jiang, Comparative electrochemical study of unsubstituted and substituted bis (phthalocyaninato) rare earth (iii) complexes. Eur. J. Inorg. Chem. **2004**, 510–517 (2004)
8. A. Abragam, B. Bleaney, Electron Paramagnetic Resonance of Transition Ions (Oxford Classic Texts in the Physical Sciences) (Oxford University Press, USA, 2012). ISBN:0199651523
9. R. Orbach, Spin-lattice relaxation in rare-earth salts. Proc. R. Soc. A: Math. Phys. Eng. Sci. **264**, 458–484 (1961)
10. J.R. Friedman, M.P. Sarachik, R. Ziolo, Macroscopic measurement of resonant magnetization tunneling in high-spin molecules. Phys. Rev. Lett. **76**, 3830–3833 (1996)
11. L. Thomas, F. Lionti, R. Ballou, D. Gatteschi, R. Sessoli, B. Barbara, Macroscopic quantum tunnelling of magnetization in a single crystal of nanomagnets. Nature **383**, 145–147 (1996)
12. W. Wernsdorfer, N. Chakov, G. Christou, Determination of the magnetic anisotropy axes of single-molecule magnets. Phys. Rev. B **70**, 1–4 (2004)
13. M. Mannini, F. Pineider, P. Sainctavit, C. Danieli, E. Otero, C. Sciancalepore, A.M. Talarico, M.-A. Arrio, A. Cornia, D. Gatteschi, R. Sessoli, Magnetic memory of a single-molecule quantum magnet wired to a gold surface. Nat. Mater. **8**, 194–197 (2009)
14. M. Urdampilleta, S. Klyatskaya, M. Ruben, W. Wernsdorfer, Landau-Zener tunneling of a single Tb $\{3+\}$ magnetic moment allowing the electronic read-out of a nuclear spin. Phys. Rev. B **87**, 195412 (2013)
15. L. Landau, Zur Theorie der Energieubertragung II. Phys. Sov. Union **2**, 46–51 (1932)
16. C. Zener, Non-adiabatic crossing of energy levels. Proc. R. Soc. A: Math. Phys. Eng. Sci. **137**, 696–702 (1932)
17. A.J. Heinrich, J.A. Gupta, C.P. Lutz, D.M. Eigler, Single atom spin-flip spectroscopy. Science (New York, N.Y.) **306**, 466–469 (2004)

Chapter 6
Nuclear Spin Dynamics—T_1

The detection and manipulation of nuclear spins has become an important multi-disciplinary tool in science, reaching from analytic chemistry, molecular biology, to medical imaging and are some of the reasons for a steady drive towards new nuclear spin based technologies. In this context, recent breakthroughs in addressing isolated nuclear spins opened up a new path towards nuclear spin based quantum information processing [1–4]. Indeed, the tiny magnetic moment of a nuclear spin is well protected from the environment, which makes it an interesting candidate for storage of quantum information [5, 6]. On this account, we are going to investigate an isolated nuclear spin using a single-molecule magnet spin-transistor in regard to its read-out fidelity and lifetime, which are important figures of merits for quantum information storage and retrieval.

6.1 Signal Analysis

The experimental results in this chapter were obtained via electrical transport measurements through a three terminal single-molecule magnet spin-transistor and by using two different samples to demonstrate the reproducibility of the data. The spin-transistors were placed into a dilution refrigerator with a base temperature of 150 mK for sample A and 40 mK for sample C. Each device was surrounded by a home-made three-dimensional vector magnet and biased at $V_{ds} = 0$ V. The gate voltage V_g was adjusted in order to shift the chemical potential of the read-out dot slightly above or below the source-drain Fermi level, resulting in the highest sensitivity of the device. From Sect. 5.1, we know that the conductance of the read-out dot at given V_{ds} and V_g depends on the direction of the electronic spin. By sweeping the external magnetic field parallel to the easy axis of the TbPc$_2$, we induced reversals of the terbium's electronic spin at the four avoided level crossings due to a quantum tunneling of magnetization (QTM). These reversal result in jumps of the read-out dot's conduc-

© Springer International Publishing Switzerland 2016
S. Thiele, *Read-Out and Coherent Manipulation of an Isolated Nuclear Spin*,
Springer Theses, DOI 10.1007/978-3-319-24058-9_6

tance. Since the magnetic field of the QTM transition is nuclear spin dependent, each conductance jump can be assigned to the nuclear spin qubit state. In the following, we will describe the data treatment in order to automatize the read-out process and explain in detail how we measured the lifetime T_1 of a single nuclear spin.

Figure 6.1a displays the raw data of five different measurements, including four sweeps where the electronic spin reversed due to a QTM transition and one sweep without a reversal. The conductance jump was evoked by a shift of the read-out dot's chemical potential due to the exchange coupling to the terbium's electronic spin (see Chap. 5).

In order to read-out the nuclear spin state, we had to analyze if, and where a conductance jump occurred during the magnetic field sweep. Therefore, the raw data were passed through a filter, which computed the first derivative with an adjustable smoothing over N data points. The output of the filtered signals from Fig. 6.1a are displayed

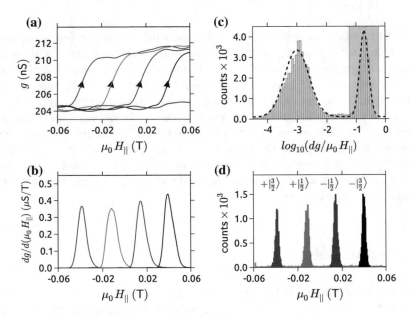

Fig. 6.1 **a** Raw data showing four measurements with a spin reversal (*blue, green, red*, and *black curve*) and one measurement without a spin reversal (*purple curve*). The conductance jump was induced by a shift of the read-out dot's chemical potential due to the exchange coupling to the electronic spin. **b** Filtered signal of (**a**), which is similar to a smoothed first derivative. Data including a reversal are transformed into peaks whose maxima indicate the respective jump position, whereas sweeps without a reversal are strongly suppressed. **c** Histogram of the maximum amplitudes of all filtered sweeps. Measurements without a spin reversal (*left peak*) can be separated from measurements containing a reversal (*right peak*) by a threshold (*yellow rectangle*). **d** Histogram of the jump positions of 75,000 measurement whose filtered maxima were within the *yellow rectangle* of (**c**). The four peaks originate from conductance jumps in the vicinity of the four anticrossings and allow for the unambiguous attribution of each detected conductance jump to a nuclear spin qubit state. The plot was generated using sample C, notice that sample A shows identical characteristics (compare Fig. 6.9a)

in Fig. 6.1b. The signal, which did not show a jump, is strongly suppressed by the filter. However, the sweeps, which contained a conductance jump, are transformed into peaks, whose maxima indicated at which magnetic field the jumps occurred.

To obtain a good statistical average, we measured the conductance signal of 75,000 magnetic field sweeps. Plotting the maximum amplitudes of all filtered data in a histogram gave rise to Fig. 6.1c. It shows that the jump amplitudes are divided into two distinct peaks, separated by more than two orders of magnitude. The left peak, corresponding to small amplitudes, originates from all measurements without a reversal; whereas the right peak, corresponding to large amplitudes, finds its origin in sweeps including a spin reversal. To sort out the measurements with spin reversals from the rest of the data we defined a threshold indicated by the yellow rectangle. If the maximum amplitude of the filtered signal lied within this rectangle, the sweep was considered to contain a QTM transition and the position of the jump was stored in an array. Subtracting the inductive field delay of the coils from the jump positions and plotting them into a histogram results in Fig. 6.1d. It shows that the conductance jumps happened almost exclusively in the vicinity of the four avoided level crossings, corresponding to the four nuclear spin states. Hence, we can unambiguously assign a nuclear qubit state to each detected jump. The width of the four peaks is determined by the lock-in time constant and the electronic noise of the setup, which leads to a broadening much larger than the intrinsic linewidth. The error induced by our nuclear spin read-out procedure is mainly due to inelastic electronic spin reversals (grey data point in Fig. 6.1), which were misinterpreted as QTM events and is estimated to be less than 5 % for sample A and less than 4 % for sample C.

6.2 Relaxation Time T_1 and Read-Out Fidelity F

After being able to read out the state of an isolated nuclear spin qubit, we are going now one step further by recording the real-time trajectory of an isolated nuclear spin. Using sample A, we present measurements, obtained by sweeping the magnetic field up and down between ±60 mT at 48 mT/s (2.5 s per sweep), while recording the conductance through the read-out quantum dot (see Fig. 6.2a). As explained in Sect. 6.1, we can assign each conductance jump to a certain nuclear spin qubit state and, due to the fixed frequency of the magnetic field ramp, to a certain time (see Fig. 6.2b).

By sweeping the magnetic field faster than the relaxation time, we obtained a real-time image of the nuclear spin trajectory. The first 2000s of this trajectory are shown in Fig. 6.3. The grey dots illustrate the position of the recorded conductance jumps. If the jump occurred within a window of ±7 mT around the avoided level crossing (indicated by colored bars), it was assigned to the corresponding nuclear spin state. If, however, a jump was recorded outside this window, the measurement was rejected. The black line shows the assigned time evolution of the nuclear spin state.

Figure 6.4a shows a magnified region of the nuclear spin trajectory including 170s of data. In order to access the nuclear spin relaxation time T_1, we performed a bit by

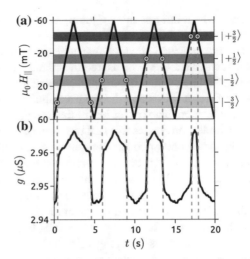

Fig. 6.2 Protocol to measure the nuclear spin trajectory. **a** The magnetic field $B_{||}$ is swept *up* and *down* between ± 60 mT at a constant rate of 48 mT/s, corresponding to 2.5 s per sweep. **b** Each detected conductance jump can therefore be assigned to a certain nuclear spin state at a certain time t

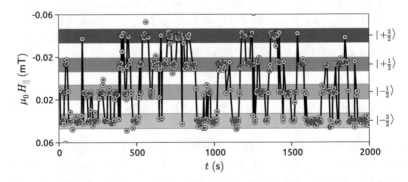

Fig. 6.3 First 2000 s of the nuclear spin trajectory measured using sample A. The *grey dots* illustrate the recorded conductance jump. If the jump was found inside a window of ± 7 mT (*colored stripes*) around one of four peaks of Fig. 6.1d it was assigned to the corresponding nuclear spin state, otherwise the measurement was rejected. Using this data analysis results in the single nuclear spin trajectory, shown as a *black line*

bit post-processing of this data. Therefore, we extracted the different dwell times, i.e. the time the nuclear spin remained in a certain state before going into another state. Plotting these dwell times for each nuclear spin state in separate renormalized histograms yielded the black data points of Fig. 6.4b–e.

A further fitting to an exponential function $y = exp(-t/T_1)$ gave the nuclear spin dependent relaxation times $T_1 \simeq 13$ s for $m_I = \pm 1/2$ and $T_1 \simeq 25$ s for $m_I = \pm 3/2$ for sample A. The perfect exponential decay indicated that no memory effect is present in the system. Furthermore, the obtained lifetimes were an order of magnitude

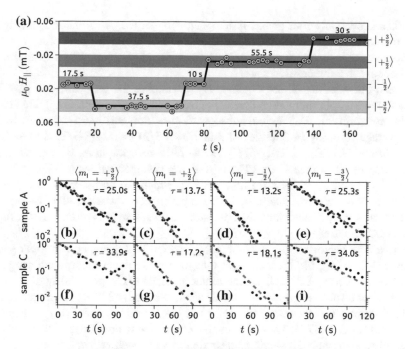

Fig. 6.4 **a** Zoom of the nuclear spin trajectory (*black curve*), which was obtained from the detected conductance jumps (*grey dots*). Every time the nuclear spin qubit changes over to a new state, we determined the dwell time in this state (*black numbers*). **b–e** Plotting the dwell times for each nuclear spin state in separate histograms led to the *black data points*. A further fitting to the exponential function $y = exp(-t/T_1)$ (*red dotted line*) yielded the relaxations times T_1 for each nuclear spin qubit state of sample A. **f–i** The relaxation times T_1 for sample C are obtained analog to sample A

larger than the measurement interval, which denotes that the same quantum state could be measured multiple times without being destroyed by the measurement process. Such a detection scheme is referred to as a quantum nondemolition (QND) read-out. Instead of demolishing the quantum system, it will only project the system onto one of its eigenstates [7]. Notice that superposition states will be destroyed by this projection.

Usually the Hamiltonian of the entire system can be written as $H = H_0 + H_M + H_I$, with H_0 being the Hamiltonian of the quantum system under study, H_M the Hamiltonian of measurement system and H_I the interaction Hamiltonian between the two systems. In order to perform a real QND measurement, it has been shown that the commutator between the measured variable q and the interaction Hamiltonian must be zero: $[q, H_I] = 0$ [7, 8]. In our experiment the measurement variable is I_z and the interaction is described by the hyperfine Hamiltonian $H_{hf} = AIJ$. The latter possesses terms of $A/2(I_+J_- + I_-J_+)$, which do not commute with I_z. This can be seen as a deviation of the ideal QND measurement. However, the Hamiltonian $A/2(I_+J_- + I_-J_+)$, accounting for flip-flop processes of the nuclear and the electronic spin represents only a weak perturbation, as it would cause additional tunnel events at all crossings in Fig. 5.8a, not marked by colored rectangles. Since

Fig. 5.8c shows only four peaks, it demonstrates that the perturbation is negligible and the deviation from an ideal QND measurement must be small. An important point to notice is that performing a QND measurement is equivalent to initialize the nuclear spin in the measured state.

The read-out fidelities F are obtained by calculating the probability to stay in a certain nuclear spin qubit state during the time necessary to measure it. Due to the QTM probability of 51.5%, two subsequent measurements were separated by ≈ 5 s in average resulting in fidelities of $F(m_I = \pm 3/2) \approx exp(-5/25.2 \text{ s}) \approx 82\%$ and $F(m_I = \pm 1/2) \approx exp(-5/13.2 \text{ s}) \approx 69\%$ for sample A. By repeating this measurement on sample C (see Fig. 6.4f–i), we obtained values of $T_1 \approx 17$ s for $m_I = \pm 1/2$ and $T_1 \approx 34$ s for $m_I = \pm 3/2$, which are comparable to sample A and shows the high reproducibility of the experiment and the excellent isolation of the nuclear spin in molecular spin-transistor devices, which is promising for future device architectures.

Due to a new vector magnet (see Sect. 4.3), which was designed for larger sweep rates (>200 mT/s), the measurement interval for the experiments with sample C could be reduced to 1.2 s. Given the rather identical QTM probability of 52% for sample C, two subsequent measurements are separated by 2.31 s in average, leading to fidelities of $F(m_I = \pm 3/2) \approx exp(-2.31/34 \text{ s}) \approx 93\%$ and $F(m_I = \pm 1/2) \approx exp(-2.3/17 \text{ s}) \approx 87\%$. These values are comparable to fidelities given by Robledo et al. [9] who measured a single nuclear spin of a nitrogen vacancy center.

The limitation of our read-out fidelity comes from the currently rather slow detection rate of 0.5 measurements per second with respect to experiments on other nuclear spin qubits, which make use of the much faster electron spin resonance (ESR). Since the magnetic field cannot be stabilized at one of the anticrossing in the Zeeman diagram, $|+6\rangle$ and $|-6\rangle$ remains the only available basis and therefore flipping the electronic spin of the TbPc$_2$ involves a $\Delta m_J = 12$. This makes the ESR process highly improbable for this system. Nevertheless, we tried to flip the electronic spin sending microwaves with the transition frequency $m_I = -1/2 \longleftrightarrow m_I = 1/2$, while sweeping the magnetic field around 0 mT. However, the expected additional transition at $B = 0$ T was not observed so far. Another possibly to perform the ESR measurements is to use the transition $m_J = 6 \rightarrow m_J = 5$. Unfortunately, the transition frequency of around 12 THz is hard to access, and our coaxial cables are not suited to guide such high frequencies.

A mid term solution would be to design special vector magnets consisting of two types of coils: larger vector magnets similar to the ones presented in Chap. 4, to generate the static magnetic field and very small coils, used to generate high frequency magnetic field ramps. In this way the detection rate could be speed up by a factor of 10–100. The long term approach, however, is to find SMMs having a strong hyperfine coupling and allowing for electronic spin transitions $\Delta m_J = \pm 1$. In an easily accessible frequency range [2 GHz $\rightarrow$ 10 GHz], these SMMs would be perfect candidates for ESR detection implementation. The resulting speed up in measurement time by 2 orders of magnitude could lead to fidelities close to 1.

6.3 Quantum Monte Carlo Simulations

In order to perform a more quantitative analysis of the nuclear spin lifetime and the involved relaxation process we wanted to make use of computational techniques. However, to do a proper quantum mechanical simulation we needed to include the coupling of the nuclear spin to a thermal bath, which requires methods that go beyond the usual solution of the Schrödinger equation. There are currently two widely used approaches to simulate such quantum trajectories. In the usual approach the master equation is written for a reduced density matrix ρ_A [10]. It computes the ensemble average of the time evolution of ρ_A. An equivalent approach is the so-called Monte Carlo wavefunction method [11–13], which calculates the stochastic evolution of the atomic wavefunction using a quantum Monte Carlo (QMC) algorithm. It can be shown that the ensemble average of the master equation is analogue to the time average of the QMC technique. However, the latter could be adapted more easily to our experimental conditions and was therefore our method of choice. The following algorithm was developed in cooperation with Markus Holzmann from the LPMCC in Grenoble.

6.3.1 Algorithm

In the following we are briefly discussing the Monte Carlo wavefunction algorithm. Notice that the complete QMC code is shown in appendix C.

Suppose the wave function of the isolated system $|\Psi\rangle$ is entirely described by the Hamiltonian H_0, and all the influence of the environment on the time evolution of the system can be described in terms of a non-Hermitian operator H_1:

$$H_1 = -\frac{i\hbar}{2} \sum_m C_m^\dagger C_m \tag{6.3.1}$$

where $C_m(C_m^\dagger)$ is an arbitrary relaxation (excitation) operator. In the following, we assume that the environment can be modeled as a bosonic bath. Furthermore, we allow only transitions of the nuclear spin, which obey $|\Delta m| = 1$, as expected from the nuclear spin transition. Thus, we get only two contributions in the Hamiltonian H_1, namely:

$$C_1^{i,j} = \sqrt{\Gamma_{i,j}(1 + n(\omega_{i,j}, T))}\, \delta_{i,j+1} \tag{6.3.2}$$

which accounts for relaxations between the state i and j, and

$$C_2^{i,j} = \sqrt{\Gamma_{i,j}(n(\omega_{i,j}, T))}\, \delta_{i+1,j} \tag{6.3.3}$$

which accounts for excitations between the state i and j in terms of their energy differences $\omega_{i,j}$ and relaxation rates $\Gamma_{i,j}$. Notice, both are symmetric in i, j and

$\omega_{i,j} = |\omega_i - \omega_j|$. Both, C_1 and C_2 have the dimension $1/\sqrt{\text{time}}$. The function $n(\Delta\omega, T) = \left(1 + exp(\frac{\hbar\Delta\omega}{k_B T})\right)^{-1}$ is the Bose-Einstein distribution, which takes the density of the bosonic bath into account; and ($\Gamma_{0,1}$, $\Gamma_{1,2}$, and $\Gamma_{2,3}$) are the state dependent transition rates, with 0, 1, 2 and 3 being the ground state, first, second, and third excited state. The effective Hamiltonian is the sum of H_0 and H_1

$$H = H_0 - \frac{i\hbar}{2} \sum_{m=1}^{2} C_m^\dagger C_m \qquad (6.3.4)$$

Notice that H_1 is non-Hermitian, since its eigenvalues are imaginary. To obtain the nuclear spin trajectory, we have to calculate the time evolution of the wavefunction, which is done in the following three steps.

Step I

In the first step we calculate the wavefunction after a small time step δt. Therefore, we make use of the classical Schrödinger equation.

$$\frac{d\tilde{\Psi}}{\delta t} = -\frac{i}{\hbar}(H_0 + H_1)\Psi$$

$$\tilde{\Psi}(t + \delta t) = exp\left(-\frac{i}{\hbar}H_1\delta t\right) exp\left(-\frac{i}{\hbar}H_0\delta t\right) \Psi(t)$$

Here, we have neglected an error of δt^2, in which case H_0 and H_1 are not commuting. Furthermore, we chose δt in a way that $\left|\frac{i}{\hbar}H_1\delta t\right| \ll 1$. Thus, the term $exp(-\frac{i}{\hbar}H_1\delta t)$ can be written in a first order Taylor series expansion $exp(-\frac{i}{\hbar}H_1\delta t) \approx 1 - \frac{i}{\hbar}H_1\delta t$. Since we are only interested in the amplitude of the wavefunction, the term $exp\left(-\frac{i}{\hbar}H_0\delta t\right)$ will be neglected in the following. It adds only a phase term to the wavefunction and can be reintroduced at any point in the calculation if necessary. Hence, the amplitude of the wavefunction after a time step δt is:

$$\tilde{\Psi}(t + \delta t) = \left(1 - \frac{i}{\hbar}H_1\delta t\right) \Psi(t) \qquad (6.3.5)$$

Step II

In the second step we calculate the transition probability from one state to another. As mentioned before the Hamiltonian H_1 is non-Hermetian and therefore the wavefunction is not normalized. Up to an error of δt^2 we can write:

$$\langle \tilde{\Psi}(t + \delta t) | \tilde{\Psi}(t + \delta t) \rangle = \langle \Psi(t) | 1 - \frac{i}{\hbar} \delta t (H_1 + H^\dagger) + O(\delta t^2) | \Psi(t) \rangle$$

$$= 1 - \delta p \tag{6.3.6}$$

with

$$\delta p = \frac{i}{\hbar} \delta t \langle \Psi(t) | \left(H_1 - H_1^\dagger \right) | \Psi(t) \rangle \tag{6.3.7}$$

Since Eq. 6.3.6 is only a first order approximation, we have to adjust δt to assure that $\delta p \ll 1$. Moreover the term δp can be written as the sum of the relaxation and excitation probability: $\delta p = \delta p_{\text{rel}} + \delta p_{\text{exc}}$, because we have only allowed those two transitions in our model, where

$$\delta p_{\text{rel}} = \delta t \langle \Psi(t) | \left(C_1^\dagger C_1 \right) | \Psi(t) \rangle$$

$$\delta p_{\text{exc}} = \delta t \langle \Psi(t) | \left(C_2^\dagger C_2 \right) | \Psi(t) \rangle$$

Step III

In the third step we will account for the random evolution of the wavefunction, which will introduce the nonreversibility of a transition. At this point the wavefunction is at a bifurcation point and could evolve in three different directions:

1 the systems stays in the same state and nothing happens,
2 a relaxation in an energetically lower state occurs,
3 the systems is excited in an energetically higher state.

In order to decide which of three events is happening, we draw and uniformly distributed pseudo-random number $\epsilon = [0, 1]$. If $\epsilon > \delta p$, no quantum jump occurs and we will renormalize the wavefunction:

$$\Psi(t + \delta t) = \frac{\tilde{\Psi}(\delta t)}{\sqrt{1 - \delta p}} \tag{6.3.8}$$

If however $\epsilon < \delta p$, the system undergoes a quantum jump. If furthermore $\epsilon < \delta p_{\text{rel}}$, we are relaxing the system according to:

$$\Psi(t + \delta t) = \frac{C_1 \tilde{\Psi}(t)}{\sqrt{\delta p_{\text{rel}}/\delta t}} \tag{6.3.9}$$

On the other hand if $\epsilon > p_{\text{rel}}$, we excite the system using the following expression:

$$\Psi(t + \delta t) = \frac{C_2 \tilde{\Psi}(t)}{\sqrt{\delta p_{\text{exc}}/\delta t}} \tag{6.3.10}$$

The denominator in Eqs. 6.3.9 and 6.3.10 accounts for the normalization of the wave-function.

6.3.2 Including the Experimental Boundaries

In order to simulate the experimentally obtained nuclear spin trajectory of Fig. 6.3 using the algorithm of Sect. 6.3, we had to introduce some slight modifications. The read-out of the nuclear spin happens due to a QTM event only once per measurement cycle and therefore at finite time steps $t_{measure}$. Furthermore, sweeping the magnetic field back and forth to measure these QTM events implicates that each nuclear spin qubit state is probed at a different time during the sweep. Moreover, the QTM transition of the electronic spin occurred with a probability of 51.5 %, and, as a consequence, reversed the order of the ground and excited states of the qubit.

To simulate this experimental conditions appropriately, the computation cycles of duration δt were grouped into five time intervals Δt_i as shown in Fig. 6.5, with $\sum_i \Delta t_i = t_{measure}$ and $t_{measure}$ being the time needed for one magnetic field sweep. The individual Δt_i were chosen in a way, that at the end of each interval, the magnetic field would have been at one of the four anticrossings corresponding to $m_1 = -3/2$, $-1/2, 1/2,$ or $3/2$ respectively. Hence, we checked every Δt_i if the nuclear spin qubit was in the appropriate state to allow for a QTM transition (question marks in Fig. 6.5). If so, we drew a second random number $\epsilon_2 = [0, 1]$ to simulate the probabilistic nature of the transition. An ϵ_2 which was smaller than the QTM probability P_{QTM}, was interpreted as a QTM event. However, an ϵ_2 that was larger than P_{QTM}, resulted in no QTM transition. Moreover, every time the QTM happened, we saved the nuclear spin state and reversed the nuclear qubit ground stated and its excited states, just like in the experiment. Once we finished the simulation of interval 5, corresponding to the end of a field sweep, we computed the time intervals in reversed order (5, 4, 3, 2, 1), which is equivalent to sweeping back the magnetic field to its initial value.

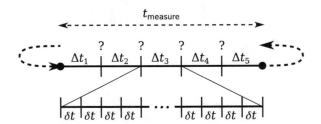

Fig. 6.5 In order to include the experimental boundaries into our simulations, the computation cycles of duration δt were grouped into intervals of Δt_i, where the sum of all Δt_i corresponds to the time needed to sweep the magnetic field during the trace or retrace measurement. At the end of each interval Δt_i, corresponding to a certain magnetic field, we checked if the nuclear spin was in the appropriate state to allow for a QTM transition. If so, the QTM event was accepted with the probability P_{QTM}, leading to the storage of the nuclear spin state and an inversion of the ground state and the excited states

6.4 Comparison Experiment—Simulation

6.4.1 Relaxation Mechanism

In the following, the computational results obtained with the algorithm of Sect. 6.3 are compared with experimental data from sample A, in order to extract further information about the underlying physics of the relaxation process. The parameters used to perform the simulation are listed in Table 6.1. The temperature T, which corresponds to the electron temperature of sample A, the measurement period of 2.5 s, and the QTM probability of 51.5 % were taken as fixed parameters. Only the transition rates Γ_{01}, Γ_{12} and Γ_{23} were varied in order to obtain the best fit to the experimental data shown in Fig. 6.6a–d.

Computing the trajectory for 2^{24} Monte Carlo time steps and following the procedure of Sect. 6.2 to extract the lifetime T_1 gave rise to the data displayed in Fig. 6.6e–f.

The first striking feature, which can be extracted from this comparison, is that the difference in T_1 between the $\pm 3/2$ states and the $\pm 1/2$ states is nicely reproduced by simulation. An explanation for this observation can be given by looking at the Hamiltonian H_1 (Eq. 6.3.1), containing the relaxation and excitation operators C_1 and C_2. If the nuclear spin is in the $|\pm 1/2\rangle$ state, the two operators C_1 and C_2 contribute to the relaxation and excitation process. If, however, the nuclear spin is in the $|\pm 3/2\rangle$ state, one of the operators becomes zero, no matter what the electronic

Table 6.1 Input parameters for the quantum Monte Carlo algorithm introduced in Sect. 6.3.

$t_{measure}$	δt	P_{QTM}	T	Γ_{01}	Γ_{12}	Γ_{23}
2.5 s	$\frac{2.5}{60}$ s	51.5 %	150 mK	1/41 s^{-1}	1/82 s^{-1}	1/90.2 s^{-1}

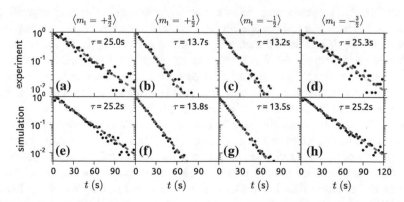

Fig. 6.6 a–d Experimental data for sample A taken from Fig. 6.4. **e–f** Computed data points using the parameters of Table 6.1 and the algorithm of Sect. 6.3. The red dotted line in each subplot is a fit to an exponential function $y = exp(-t/T_1)$, yielding the relaxation time T_1 for each nuclear spin state

spin state is, resulting in a smaller transition rate and therefore a larger T_1. A more descriptive explanation can be given by considering the number of transition paths. If the nuclear spin is in the ground or most excited state ($m_1 = \pm 3/2$), there is only one way to change its state—excitation **or** relaxation, whereas if the nuclear spin is in an intermediate state ($m_1 = \pm 1/2$) it has two escape paths—excitation **and** relaxation. Since the lifetime is roughly inversely proportional to the number of transition paths, if the rates for each part were equal, the T_1's show a difference of approximately two. The exact ratio depends of course on the temperature and the individual transition rates.

In the next step we wanted to reveal the dominant relaxation mechanism, which could be caused by spin-lattice interactions and nuclear spin diffusion. The latter mechanism was found to be very weak in bulk terbium [14] and can, hence, be neglected for rather isolated and nonaligned SMMs. Concerning the spin-lattice relaxation mechanism, we examined closer the $\Gamma_{i,j}$'s derived by fitting the results of QMC simulations to experimental data. Depending on its proportionality to the nuclear level spacing $\omega_{i,j}$ we can distinguish between three types of mechanisms.

1 The Korringa process, in which conduction electrons polarize the inner lying s-electrons. Since these couple with the nuclear spins via contact interaction, an energy exchange over this interaction chain is established, leading to $\Gamma_{i,j} \propto |\langle i|I_x|j\rangle|^2$ [15].

2 The Weger process, which suggests that the spin-lattice relaxation is dominated by the intra-ionic hyperfine interaction and the conduction electron exchange interaction [16]. It is a two-stage process, where the energy of the nucleus is transmitted to the conduction electrons via the creation and annihilation of a Stoner excitation. This process is similar to the Korringa process but results in $\Gamma_{i,j} \propto |\langle i|I_x|j\rangle|^2 \omega_{i,j}^2$.

3 The magneto-elastic process, which leads to a deformation of the molecule due to a nuclear spin relaxation, yields $\rightarrow \Gamma_{i,j} \propto |\langle i|I_x|j\rangle|^2 \omega_{i,j}^4$ [17].

The term $|\langle i|I_x|j\rangle|^2$ arises from the fact, that only rotations of the spin perpendicular to the z-directions are responsible for longitudinal transitions [18]. A comparison between the $\Gamma_{i,j}$'s and the different mechanisms is shown in (Fig. 6.7a). The almost perfect agreement with the Weger process suggests that the dominant relaxation process is caused by the conduction electrons. Since they are exchange coupled to the Tb electronic spin which in turn is hyperfine coupled to the nuclear spin, an energy and momentum exchange via Stoner excitations could be possible.

This implies that by controlling the amount of available conduction electrons per unit time the relaxation rates $\Gamma_{i,j}$ can be changed. Hence, an electrically control of T_1 by means of the bias and gate voltages is possible. To verify this conclusion, we measured the relaxation time T_1 of $m_1 = \pm 3/2$ as a function of the tunnel current through the quantum dot. The result in Fig. 6.7b shows a decrease of the lifetime by a factor of three while increasing the current by 100 %. This finding could be interesting to speed up the initialization of the nuclear spin in its ground state prior to a quantum operation by inducing a fast relaxation to the ground state through a series of current pulses.

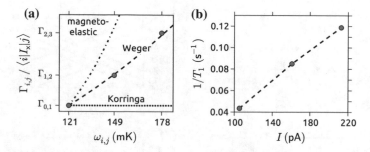

Fig. 6.7 **a** The transition rates $\Gamma_{i,j}$, derived by fitting the results of QMC simulations to experimental data, exhibit a quadratic dependence on the nuclear spin level spacing $\omega_{i,j}$. This behavior is expected from a Weger relaxation process, in which the nuclear spin is coupled via virtual spin waves to conduction electrons. **b** The decrease of lifetime with increasing current is probably due to an increase of electrons tunneling through the read-out dot in addition to an increase of temperature

Another experiment which shows the coupling of the nuclear spin to the electrons was carried out by measuring the nuclear spin temperature as a function of the applied bias voltage. Since we are dealing with a single nuclear spin, the physical quantity temperature has only a meaning if we are speaking of time averages.

The read-out fidelity of the nuclear spin state is rapidly decreasing for increasing bias voltages as shown in Fig. 5.11a. Therefore we developed the protocol shown in Fig. 6.8a in order to measure the nuclear spin at bias voltages beyond 300 μV.

In a first step, the source-drain voltage V_{ds} was rapidly increased from 0 V to values of 1 and 2 mV, at which we waited for 6 s. The tunnel current through the quantum dot at 1 mV was about 1 nA. Extrapolating Fig. 6.7 to 1 nA results in a T_1 of 1.49 s, which was four times smaller than the waiting time and therefore long enough to thermalize the nuclear spin. During this time, the magnetic field was at -60 mT, leading to ground state of $m_I = -3/2$ (see Fig. 6.8b).

Afterward, V_{ds} was decreased to 0 V in order to probe the nuclear spin state by sweeping the magnetic field to $+60$ mT and back while checking for a QTM transition. Repeating this procedure 6000 times for 0, 1, and 2 mV led to the black histograms of Fig. 6.8c–e, showing the four peaks corresponding to the four nuclear spin states. Integrating each peak over a window of ± 7 mT around the maximum and normalizing the outcome led to the nuclear spin population, which was subsequently fitted to the Boltzmann distribution (red dotted line in Fig. 6.8c–e). From the fitting parameters, we obtained the time average nuclear spin temperature.

As shown in Fig. 6.8f, the temperature is increasing monotonically with augmenting V_{ds}, which demonstrates the coupling of the nuclear spin to the electronic bath. During the experiment the temperature of the cryostat was stable at 150 mK, suggesting that the increase of the time average nuclear spin temperature is caused by an energy exchange with the electrons tunneling through the read-out quantum dot. A deeper analysis, however, is quite difficult since the local Joule heating of the device is unknown. More insights to this topic might be provided by Clemens Winkelmann et al., working at the Néel institute. They started a three year project, dedicated to investigate the heat conduction through a single molecule inside a breakjunction using local thermometers.

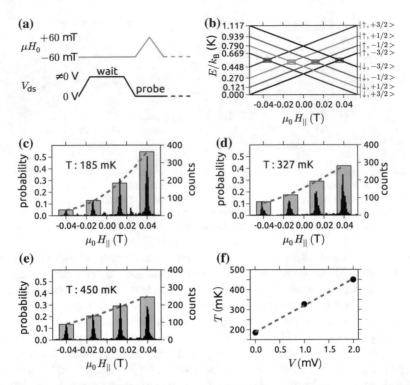

Fig. 6.8 **a** Protocol to measure the time average nuclear spin temperature. The source-drain voltage V_{ds} is rapidly increased to a finite value, at which we thermalized the nuclear spin. Afterward, V_{ds} is brought by to zero in order to probe the nuclear spin state by sweeping the magnetic field from −60 to 60 mT and back. **b** Zeeman diagram of the nuclear spin. During the waiting period in (**a**) the external magnetic field is at −60 mT, making $m_I = -3/2$ the ground state of the nuclear qubit. **c–e** Histogram of 6000 sweeps at 0 mV (**c**), 1 mV (**d**) and 2 mV (**e**) V_{ds} offset. The *grey bars* show the time average population of the nuclear spin and were obtained by integrating each peak over a window of ±7 mT around its maximum. Fitting the population to a Boltzmann distribution (*red dotted curve*) allowed for the extraction of the time average nuclear spin temperature T. **f** Fitted temperatures of (**c–e**) versus the applied source-drain voltage V_{ds}

6.4.2   Dynamical Equilibrium

Measuring the nuclear spin trajectory by sweeping the magnetic field up and down leads to an inversion of the nuclear qubit ground state and the excited states at every QTM transition. Since this inversion period is smaller than T_1, the time-average population of the nuclear spin converges to a dynamical equilibrium, which is far from the thermal Boltzmann distribution. Plotting the data obtained from nuclear spin trajectory in a histogram, and integrating over each of the four peaks, reveals the average population within this dynamical equilibrium (see Fig. 6.9a). It shows that the probability for being in each state is not 25 %, but slightly larger for $m_I = \pm 1/2$ compared to $m_I = \pm 3/2$. The time-average population obtained by the QMC

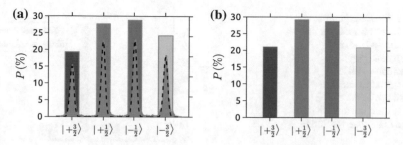

Fig. 6.9 **a** Histogram of the data obtain during the measurement of the nuclear spin trajectory of sample A. The *grey bars* correspond to the integral over each peak, revealing the time-average population of each nuclear spin. **b** Time-average population simulated using the parameters of Table 6.1 and the algorithm of Sect. 6.3. The higher probability of $m_I = \pm 1/2$ with respect to $m_I = \pm 3/2$ comes from the difference in the transition rates $\Gamma_{0,1}$ and $\Gamma_{2,3}$, and the periodical inversion of the ground state and the most excited states due to a QTM transition. For more details see text

simulations shows the same feature (see Fig. 6.9b), which allows for an explanation within the framework of the QMC model.

We found that the shape of the time-average population, in the case where the measurement time t_{measure} is smaller than T_1, is mainly governed by the individual transition rates $\Gamma_{i,j}$. As shown in Table 6.1, $\Gamma_{0,1}$ is much smaller than $\Gamma_{2,3}$, which causes a faster transition from the most excited state into the second excited state than from the first excited state into the ground state. Due to this asymmetry, and the periodic inversion of the ground state and the excited states, we are actively pumping the population into $m_I = \pm 1/2$ states. Notice that for equal $\Gamma_{i,j}$'s the time-average population would be 25 % for each state.

6.4.3 Selection Rules

During the analysis of the nuclear spin trajectory, we observed transitions with $\Delta m_I \neq \pm 1$. In order to clarify if this effect arose from a finite time resolution, i.e. multiple $\Delta m_I = \pm 1$ transitions between two subsequent measurements or additional transition paths, allowing for $\Delta m_I \neq \pm 1$, we compared experimental and simulated data. By counting the number of transitions corresponding to $\Delta m = 0$, ± 1, ± 2 and ± 3 and normalizing them with respect to the total amount of transitions, we obtained the red histogram in Fig. 6.10. Repeating this protocol for the simulated nuclear spin trajectory gave rise to the grey histogram.

The good agreement with the experimental data supports our assumption that the nuclear spin can only perform quantum jumps, which change its quantum number by one since the computational model allowed only for such transitions. All higher orders of Δm_I are therefore multiple transitions of $\Delta m_I = \pm 1$, which were not resolved due to the finite time resolution.

Fig. 6.10 Histogram of all transitions observed in the experiment (*red*) and simulation (*grey*). Transitions with $\Delta m_1 \neq \pm 1$ correspond to multiple transitions of $\Delta m_1 = \pm 1$, which were not resolved due to the finite time resolution

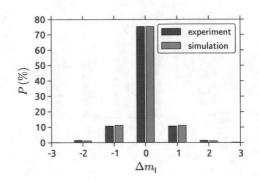

6.5   Summary

In this chapter we presented the dynamical evolution of the nuclear spin. Making use of the single-molecule magnet spin-transistor as a detection device, we recorded the real-time nuclear spin qubit trajectory over many days. Using a post treatment of the experimental data, we could extract the relaxation time T_1 for each nuclear spin state individually. Repeating this measurement on a second sample confirmed that the lifetime T_1 was in the order of a few tens of seconds, showing that the nuclear spin is well protected in the our devices. In order to perform a more sophisticated analysis of the experimental data, we developed a quantum Monte-Carlo code to numerically retrace the nuclear spin evolution. Fitting the simulation to the experimental data led to the extraction of the otherwise hardly accessible state dependent relaxation rates of the nuclear spin. These were found to depend strongly on type of relaxation, which enabled us to identify that the nuclear spin relaxation is dominated by an energy exchange with the electrons tunneling through the read-out quantum dot. An experimental confirmation of this conclusion was found in the tunabilty of the nuclear spin lifetime T_1 with respect to the tunnel-current. Additional evidence of the coupling between the nuclear spin and the tunnel electrons could be found through an increase of the nuclear spin temperature with augmenting tunnel current. Moreover, the experiments shed light on the read-out fidelities of the nuclear qubit, which were better than 69 and 87 % for sample A and C respectively, and are important figures of merit toward single-molecule magnet based quantum bits.

References

1. M.V.G. Dutt, L. Childress, L. Jiang, E. Togan, J. Maze, F. Jelezko, A.S. Zibrov, P.R. Hemmer, M.D. Lukin, Quantum register based on individual electronic and nuclear spin qubits in diamond. Science (New York, N.Y.) **316**, 1312–1316 (2007)
2. P. Neumann, J. Beck, M. Steiner, F. Rempp, H. Fedder, P.R. Hemmer, J. Wrachtrup, F. Jelezko, Single-shot readout of a single nuclear spin. Science (New York, N.Y.) **329**, 542–544 (2010)

3. R. Vincent, S. Klyatskaya, M. Ruben, W. Wernsdorfer, F. Balestro, Electronic read-out of a single nuclear spin using a molecular spin transistor. Nature **488**, 357–360 (2012)
4. J.J. Pla, K.Y. Tan, J.P. Dehollain, W.H. Lim, J.J.L. Morton, F.A. Zwanenburg, D.N. Jamieson, A.S. Dzurak, A. Morello, High-fidelity readout and control of a nuclear spin qubit in silicon. Nature **496**, 334–338 (2013)
5. B.E. Kane, A silicon-based nuclear spin quantum computer. Nature **393**, 133–137 (1998)
6. P.C. Maurer, G. Kucsko, C. Latta, L. Jiang, N.Y. Yao, S.D. Bennett, F. Pastawski, D. Hunger, N. Chisholm, M. Markham, D.J. Twitchen, J.I. Cirac, M.D. Lukin, Room-temperature quantum bit memory exceeding one second. Science (New York, N.Y.) **336**, 1283–1286 (2012)
7. V. Braginsky, F. Khalili, Quantum nondemolition measurements: the route from toys to tools. Rev. Mod. Phys. **68**, 1–11 (1996)
8. V.B. Braginsky, Y.I. Vorontsov, K.S. Thorne, Quantum nondemolition measurements. Science (New York, N.Y.) **209**, 547–557 (1980)
9. L. Robledo, L. Childress, H. Bernien, B. Hensen, P.F.A. Alkemade, R. Hanson, High-fidelity projective read-out of a solid-state spin quantum register. Nature **477**, 574–578 (2011)
10. C. Cohen-Tannoudji, Frontiers in laser spectroscopy. in *Les Houches Summer School Proceedings* (1975)
11. J. Dalibard, Y. Castin, K. Mølmer, Wave-function approach to dissipative processes in quantum optics. Phys. Rev. Lett. **68**, 580–583 (1992)
12. K. Mølmer, Y. Castin, J. Dalibard, Monte Carlo wave-function method in quantum optics. J. Opt. Soc. Am. B **10**, 524 (1993)
13. K. Mølmer, Y. Castin, Monte Carlo wavefunctions in quantum optics. Quantum Semiclassical Opt. J. Eur. Opt. Soc. Part B **8**, 49–72 (1996)
14. N. Sano, J. Itoh, Nuclear magnetic resonance and relaxation of 159 Tb in ferromagnetic terbium metal. J. Phys. Soc. Jpn. **32**, 95–103 (1972)
15. J. Korringa, Nuclear magnetic relaxation and resonnance line shift in metals. Physica **16**, 601–610 (1950)
16. M. Weger, Longitudinal nuclear magnetic relaxation in ferromagnetic iron, cobalt, and nickel. Phys. Rev. **128**, 1505–1511 (1962)
17. N. Sano, S.-I. Kobayashi, J. Itoh, Nuclear magnetic resonance and relaxation of Dy 163 in ferromagnetic dysprosium metal at low temperature. Prog. Theoret. Phys. Suppl. **46**, 84–112 (1970)
18. M. McCausland, I. Mackenzie, Nuclear magnetic resonance in rare earth metals. Adv. Phys. **28**, 305–456 (1979)

Chapter 7
Nuclear Spin Dynamics—T_2^*

Nuclear spin qubits are interesting candidates for quantum information storage due to their intrinsically long coherence times. In the last chapter, we investigated the relaxation time T_1 of a single nuclear spin and the read-out fidelity. Thus, having demonstrated four out of five DiVincenzo criteria, we turn now to the coherent manipulation of the nuclear qubit, which will complete the list.

To perform such a manipulation on a nuclear spin, large resonant AC magnetic fields are necessary. To be able to address spins individually, those AC fields are usually generated by driving large currents through nearby microcoils [1]. Yet, in order to reduce the parasitic cross talk to the read-out quantum dot and the Joule heating of the device, the maximum amplitude of the magnetic field is limited and rarely exceeds a few mT [2].

To avoid those problems, especially the Joule heating, a manipulation by means of an electric field is advantageous, in particular for scalable device architectures. Since the electric field is unable to rotate the nuclear spin directly, an intermediate quantum mechanical interaction is necessary, which transforms the electric field into an effective magnetic field. Such interactions are for example the spin-orbit coupling [3, 4], the g-factor modulation [5], or the hyperfine interaction [6].

In this chapter we will show how the latter can be used to perform coherent rotations of the nuclear spin, which are up to two orders of magnitude faster than state of the art micro-coil approaches.

7.1 Introduction

7.1.1 Rabi Oscillations

Any two level spin qubit system is characterized by its two spin orientations $|\uparrow\rangle$ and $|\downarrow\rangle$. To visualize such a system, people make use of the Bloch sphere representation.

© Springer International Publishing Switzerland 2016
S. Thiele, *Read-Out and Coherent Manipulation of an Isolated Nuclear Spin*,
Springer Theses, DOI 10.1007/978-3-319-24058-9_7

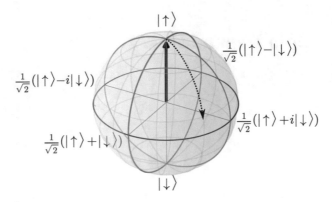

Fig. 7.1 Bloch *sphere* representation of a two level spin qubit system. The north and south pole of the sphere correspond to the two eigenstates $|\uparrow\rangle$ and $|\downarrow\rangle$, whereas the qubit state is indicated as a vector, which can be at any point on the surface. A coherent manipulation of the qubit is shown as a rotation of the vector on the sphere. Note that the trajectory of the rotation was chosen arbitrary and has no further meaning

Therein, the qubit state is symbolized as a Bloch vector, pointing from the origin of the sphere towards its surface. Moreover, the two eigenstates $|\uparrow\rangle$ and $|\downarrow\rangle$ correspond to the north and south pole of the sphere and any linear superposition $a|\uparrow\rangle + b|\downarrow\rangle$ is depicted as a point on the sphere's surface. To complete the picture, any coherent manipulation of the qubit can be illustrated as a rotation of the Bloch vector around the sphere (Fig. 7.1).

In order to manipulate the two level spin qubit, we first have to lift its degeneracy. This can be done by applying a static magnetic field B_z along the z-axis. The Hamiltonian accounting for this effect is the Zeeman Hamiltonian (see Sect. 3.3):

$$H_Z = \hbar\omega_z\sigma_z \tag{7.1.1}$$

with $\hbar\omega_z = g\mu B_z$ being the separation between the ground state and the first excited state and σ_z the Pauli spin operator, which performs a quantum mechanical operation that can be thought of a precession of the spin around the z-axis.

Now, the actual manipulation of the spin qubit requires an AC magnetic field in x- or y-direction. Without loss of generality, we assume that the magnetic field of magnitude $2B_1$ is applied along the x-axis. Decomposing the term into two counter-rotating parts as shown in Fig. 7.2 will simplify the calculation.

$$\boldsymbol{B}_R = B_1 \left(cos(\omega t)\boldsymbol{e}_x + sin(\omega t)\boldsymbol{e}_y \right) \tag{7.1.2}$$

$$\boldsymbol{B}_L = B_1 \left(cos(\omega t)\boldsymbol{e}_x - sin(\omega t)\boldsymbol{e}_y \right) \tag{7.1.3}$$

Furthermore, we assume $\boldsymbol{B}_R$ will rotate in sense with the nuclear spin precession and $\boldsymbol{B}_L$ in the opposite sense. In the frame work of the rotating wave approximation,

Fig. 7.2 Decomposition of
the AC magnetic field
$B = 2B_1cos(\omega t)e_x$ into two
counter-rotating parts

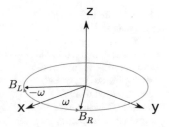

one can show that near the resonance ($\omega \simeq \omega_z$), the counter-rotating part can be neglected [7] and the time dependent part becomes:

$$H_{AC} = \hbar\Omega \left(cos(\omega t)\sigma_x + sin(\omega t)\sigma_y\right) \qquad (7.1.4)$$

with $\hbar\Omega = g\mu B_1$ and σ_x and σ_y are the Pauli spin matrices, accounting for rotations around x and y. The qubit Hamiltonian H, including both contributions $H_Z + H_{AC}$, is given as:

$$H = \hbar\omega_z\sigma_z + \hbar\Omega \left(\sigma_x cos(\omega t) + \sigma_y sin(\omega t)\right) \qquad (7.1.5)$$

To simplify the equation the following equality is applied [7]:

$$\sigma_x cos(\omega t) + \sigma_y sin(\omega t) = e^{-i\omega t\sigma_z}\sigma_x e^{i\omega t\sigma_z} \qquad (7.1.6)$$

resulting in:

$$H = \hbar\omega_z\sigma_z + \hbar\Omega e^{-i\omega t\sigma_z}\sigma_x e^{i\omega t\sigma_z} \qquad (7.1.7)$$

In order to eliminate the phase factors $e^{\pm i\omega t\sigma_z}$, we perform a unitary transformation of $U = exp(i\omega t\sigma_z)$. Physically, this can be understood as switching from the laboratory frame to the frame rotating around the z-axis with the frequency ω. To write the Hamiltonian in its usual way, we introduce $\Delta = \omega_z - \omega$, being the detuning between the MW frequency and the qubit level spacing.

$$H = \frac{\hbar\Delta}{2}\sigma_z + \frac{\hbar\Omega}{2}\sigma_x \qquad (7.1.8)$$

To visualize the enormous advantage of the rotating frame approximation, we calculated the evolution of the qubit wavefunction exposed to an AC magnetic field in x-direction in the laboratory frame (see Fig. 7.3a) and in the rotating frame (see Fig. 7.3b). We assumed that the qubit was at $t = 0$ in the $|\uparrow\rangle$ state (grey vector). To compute the trajectory on the Bloch sphere, we used the Qutip [8, 9] master equation solver. Therein, the wavefunction $|\Psi\rangle = a|\uparrow\rangle + +b|\downarrow\rangle$ is calculated at different times steps, in which the expectation values σ_x, σ_y, and σ_z were evaluated. The Python code using the Qutip library to generate Fig. 7.3 is presented in appendix D.

Fig. 7.3 The trajectory of a spin qubit, initialized in the $| \uparrow \rangle$ state (*grey arrow*) at $t = 0$, was computed in the laboratory frame (**a**) and the rotating frame (**b**), while being exposed to an AC magnetic field in x-direction at resonance frequency ($\Delta = 0$)

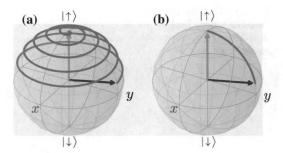

Note that in the rotating frame, the magnetic field in z-direction is proportional to Δ and therefore zero at the resonance frequency, whereas it is B_z in the laboratory frame. The big advantage of the rotating frame is that all fields are static, which allows for an easy superposition of the different components. Hence, at $\Delta \neq 0$, the Bloch vector rotates around a vector of angle $\theta = arctan(\Omega/\Delta)$ with respect to the z-axis (see Fig. 7.4), and the frequency of the precession is simply given by:

$$\Omega_R = \sqrt{\Delta^2 + \Omega^2} \tag{7.1.9}$$

with Ω_R being the Rabi frequency.

To actually measure the precession trajectory on the Bloch sphere, as presented in Fig. 7.5a, MW pulses with different duration τ are applied (see Fig. 7.5b). Before each pulse, the qubit is initialized in the $| \uparrow \rangle$ state. The following pulse is rotating the spin with the frequency Ω_R around an axis given by Ω and Δ. After the duration τ, the expectation value of σ_z of the qubit is measured. By plotting the expectation value versus the pulse duration τ, we obtain Rabi oscillations as shown in Fig. 7.5c. The amplitude and the frequency of the oscillations strongly depends on the detuning and power of the MW. Note that the largest amplitude is found at the resonance.

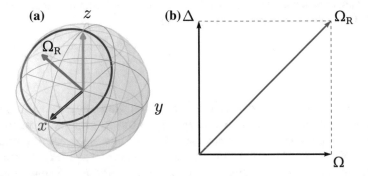

Fig. 7.4 **a** Spin precession around an effective magnetic field in the rotating frame. **b** The precession frequency Ω_R is given by $\Omega_R = \sqrt{\Delta^2 + \Omega^2}$

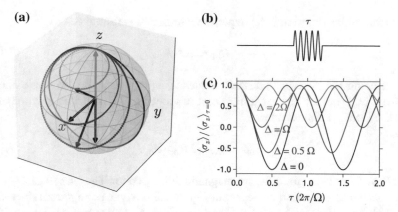

Fig. 7.5 a Trajectory of the Bloch vector in the rotating frame at different detunings $\Delta = 0$ (*black curve*), $\Delta = 0.5\Omega$ (*blue curve*), $\Delta = \Omega$ (*red curve*), and $\Delta = 2\Omega$ (*blue curve*). The spin was initialized in the ground state (*grey vector*) and exposed to pulses with different duration τ (**b**). The expectation value $\langle \sigma \rangle_z$ was evaluated at the different pulse durations and different detunings resulting in the Rabi oscillations (**c**) whose amplitude is largest at the resonance frequency ($\Delta = 0$)

Moreover, the Rabi frequency at $\Delta = 0$ is $\Omega_R = g\mu B_1/\hbar$. For an electronic spin μ is the Bohr magneton, however, for the nuclear spin $\mu = \mu_N$, the nuclear magneton, which is 2000 times smaller than μ_B. Hence, to manipulate a nuclear spin with the same speed as an electron spin, three orders of magnitude larger magnetic fields are necessary. The usual approach to generate AC magnetic makes use of on-chip microcoils, which are in the vicinity of the qubit. Yet, the parasitic cross talk to the quantum dot and the Joule heating of the entire sample limit the magnetic fields to a few mT [2]. To circumvent these problems, a manipulation could be performed by means of an electric field. Especially the Joule heating is tremendously reduced, which is of major importance for scalable device architectures. Since the electric field is not able to rotate the spin directly, an intermediate quantum mechanical interaction is necessary to transform the electric field into an effective magnetic field. Such interactions are for example the spin-orbit coupling [3, 4], the g-factor modulation [5], or the hyperfine interaction [6]. In order to manipulate a nuclear spin, the latter seems the most suited and will therefore be in the focus of the next chapter.

7.1.2 Hyperfine Stark Effect

The origin of the hyperfine coupling is exlained as a dipolar coupling between the nuclear magnetic moment μ_I, and the orbital magnetic moment μ_L and spin magnetic moment μ_S of the electron respectively. Notice, there exsists a second, although smaller contribution, which origins from the nonzero probability density of s-electrons at the core. It is referred to as the Fermi contact interaction and only relevant for s-shell electrons.

The Hamiltonian describing the hyperfine interaction is formulated as:

$$H_{hf} = A I J \tag{7.1.10}$$

with A being the hyperfine constant, I nuclear spin, and J the electronic angular momentum. From the nuclear spin's point of view, Eq. 7.1.10 can be rewritten as an effective Zeeman Hamiltonian:

$$H_{hf} = g_N \mu_N I B_{eff}(A, J) \tag{7.1.11}$$

with $B_{eff}(A, J)$ being the effective magnetic field operator. The terms $I_+ J_- + I_- J_+$, which account for electron-nuclear spin-flip transitions, can be neglected since $m_J = \pm 5$ levels are separated by 600 K. Therefore $B_{eff}(A, J)$ can be associated with an ordinary magnetic field at the center of the nucleus.

In order to create an effective AC magnetic field, $B_{eff}(A, J)$ needs to be modulated periodically. If this modulation is done by means of an electric field, we referred to it as the hyperfine Stark effect. In analogy to the ordinary Stark effect, which describes the modification of the electronic levels under an external electric field, the hyperfine Stark effect deals with the shift of the nuclear energy levels.

One of the first experimental evidence of this effect was given by Haun et al. [10]. They investigated the shift of the hyperfine transition $|F = 4, m_F = 0\rangle \longleftrightarrow |F = 3, m_F = 0\rangle$ for the ^{133}Cs ground state (see Fig. 7.6). In their measurements they observed a quadratic dependence of the level splitting on the electric field. Since the level shift remains small compared to the hyperfine splitting, an explanation of this behavior can be given by first order perturbation theory. If e is the electric charge of the electron, $\mathcal{E}$ the electric field, and r component of the vector connecting the nucleus and the electron along $\mathcal{E}$, the perturbation is given by $er\mathcal{E}$. Since this is a odd-parity term and the atomic ^{133}Cs ground states are of well defined parity, all first order perturbation terms are zero. The first nonzero elements occur in second order of perturbation and contain $(er\mathcal{E})^2$, which gives rise to the quadratic Stark shift.

Fig. 7.6 Shift of the $F = 4, m_F = 0 \longleftrightarrow F = 3, m_F = 0$ transition frequency of the ^{133}Cs ground state as a function of the square of the applied voltage

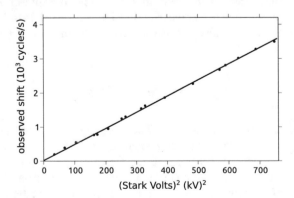

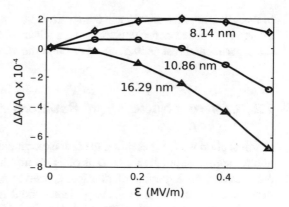

Fig. 7.7 Electric field response of the hyperfine constant at different distances between the impurity and the silicon interface. Adapted from [12]

In 1998, Kane applied the idea of the hyperfine Stark effect on ^{31}P nuclear spin qubits in silicon. He suggested that by using local gates at each qubit, the different nuclear spins can be tuned in and out of resonance independently [11]. This way, he established the individual addressability of nuclear spin qubits unsing only a global microwave field.

To show the feasibility of Kane's idea, Rahman et al. evaluated the hyperfine Stark shift of a ^{31}P impurity near the silicon interface withing the framework of the tight binding theory [12]. Since the interface breaks the symmetry around impurity, the wavefunctions of the ^{31}P are modified, resulting in states with mixed parity. Therefore, the first order perturbation terms are nonzero, giving rise to a change of the hyperfine splitting which is linear in $\mathcal{E}$. In their model, this modification is expressed as a change of the hyperfine constant $\Delta A/A_0$, which was found to be up to $\approx 10^{-3}$ at electric fields of 1 MV/m (see Fig. 7.7).

Now we turn to the TbPc$_2$ SMM. From Sect. 3.7 we know that the hyperfine constant of the Tb^{3+} inside the molecule is $A = 24.9$ mK [13]. Using Eqs. 7.1.10 and 7.1.12 we obtain an effective magnetic field at the nucleus of:

$$\boldsymbol{B}_{\text{eff}}(A, J) = \frac{AJ}{g_N \mu_N} = 313\ T \qquad (7.1.12)$$

which is two orders of magnitude larger than the usual laboratory fields. Assuming we could periodically modify the hyperfine constant A by 1/1000, we would be able to generate AC magnetic field of ± 313 mT. Since the orientation of the quantization axis of the molecule with respect to the electric field is not well determined, the effective magnetic field will have components in the x- and z-direction. However, in terms of oscillating fields only the component in x-direction is able to rotate the nuclear spin, whereas the z-component induces additional decoherence. Moreover, we can predict a linear response to an external magnetic field, since the phthalocyanine ligands break the inversion symmetry of the Tb^{3+}, analog to the ^{31}P impurities at the interface. Therefore the first, instead of the second harmonic, of the oscillating electric field must be matched to the nuclear transition frequency.

In the following sections we will demonstrate how we used the hyperfine Stark effect to perform a coherent manipulation of a nuclear spin. Additionally, we will compare the experimental results to a more profound theoretical model.

7.2 Coherent Nuclear Spin Rotations

In this section we are presenting the first experimental evidence of a coherent single nuclear spin manipulation by means of an electric field. As pointed out in the previous sections, the hyperfine Stark effect is used as a mediating quantum mechanical process to transform an oscillating electric field into an AC magnetic field. This procedure can be viewed as the AC extension of Kane's proposal form 1998, and will allow for the generation of large amplitude local magnetic fields without the inconvenience of using large AC currents through close by microcoils.

In order to simplify the problem, we will focus on the nuclear spin subspace containing only the $|+3/2\rangle$ and $|+1/2\rangle$ qubit states. By assigning the $|+3/2\rangle$ and $|+1/2\rangle$ states the Bloch vectors pointing to the north and south pole of the Bloch sphere respectively, we can use the theory that was presented in Sect. 7.1.1 to explain the quantum manipulation. However, in this subspace the operator I_x becomes $\sqrt{3}\sigma_x$, I_y becomes $\sqrt{3}\sigma_y$, and I_z becomes σ_z, with $\sigma_{x,y,z}$ being the corresponding Pauli spin 1/2 matrices. Note that the other two nuclear spin subspaces would have worked as well.

7.2.1 Frequency Calibration

The coherent manipulation of the nuclear spin qubit requires the knowledge of the exact level spacing between the $|+3/2\rangle$ and $|+1/2\rangle$ states. This frequency depends of course on the electrostatic environment due to the hyperfine stark effect. A first indication of the approximate position of the resonance frequency could be found in the work of Hutchison and Ishikawa [13, 14] who gave values of 2.3 and 2.5 GHz.

In conventional NMR experiments, the nuclear spins start to absorb a notable amount of microwave power at the resonance frequency, which can be detected by a change of the reflection or transmission of the microwave signal. In case of a single nuclear spin, this signal is much to small to be detected. Therefore, we developed our own protocol, which is sensitive to an increase of the relaxation rate if the two nuclear spin transition is in resonance to the frequency of the applied AC electric field.

The schematic of the protocol is shown in Fig. 7.8a. First, the nuclear spin was initialized by sweeping the magnetic field $\mu_0 H_\parallel$ from negative to positive values (purple curve) while checking for a QTM transition at one of the 4 avoided level crossings (see colored rectangles (b)). Subsequently, we applied a MW pulse of duration $\tau = 1$ ms. The final state is then detected by sweeping back the external

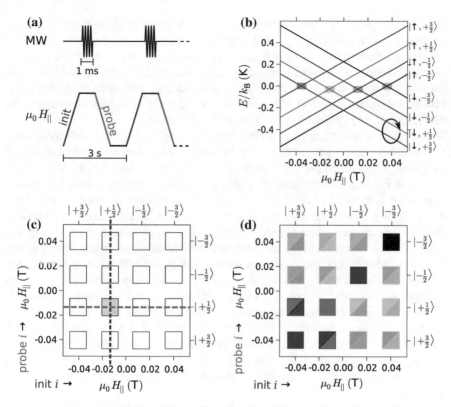

Fig. 7.8   a Measurement protocol to find the resonance frequency. To initialize the nuclear spin, the magnetic field $\mu_0 H_\parallel$ is swept from negative to positive values (*purple curve*), while checking for QTM transition. Subsequently, we kept $H_\parallel$ constant (*black curve*) and applied a microwave (MW) pulse of 1 ms. In the end, the final state is probed by sweeping back $H_\parallel$ to negative values (*orange curve*). One measurement cycle has a duration of 3 s and is therefore much faster than T_1. **b** If the microwave was in resonance with the two lowest nuclear qubit levels, a transitions between $m_I = +3/2 \longleftrightarrow m_I = +1/2$ could be induced at positive $H_\parallel$. **c** Schematic showing the construction of a 2D matrix to visualize the transitions. In the shown example the nuclear spin was initialized in the $|+1/2\rangle$ state (*vertical line*) and probed in the same state (*horizontal line*) giving to an element on the diagonal line. **d** Full 2D matrix. The elements on the diagonal, having only one color, correspond to measurements, where the nuclear spin state was not change between the initialization an the probe sweep. If the microwave was in resonance with the $m_I = +3/2 \longleftrightarrow m_I = +1/2$ transition, increased offdiagonal will appear, as indicated by blue-green rectangles. The other offdiagonals will also appear due to relaxation processes but with much less intensity

magnetic field in a time scale faster than the measured relaxation times of both nuclear spin states. The entire sequence is rejected when the initial or final state was not detected due to a missing QTM transition. A full cycle had a duration of 3 s and was therefore much faster than T_1 so that thermal relaxation processes were observed only every 6–11 measurements, depending on the nuclear spin state.

The MW was pulsed because of less heating of the device with respect to a continuous irradiation. However, the pulse width should be larger than the dephasing time T_2^*, which was expected to be smaller than 1 ms, in order to avoid accidental full coherent rotations in the Bloch sphere, which preserve the nuclear spin state. If the MW frequency was in resonance with the two lowest nuclear qubit levels at positive $H_{||}$, a transition between $m_I = +3/2 \longleftrightarrow m_I = +1/2$ could be induced (see Fig. 7.8b) resulting in an increased relaxation rate of the two states.

To visualize the relaxation rate, we constructed a two-dimensional matrix as follows. The detected nuclear spin state during the initialization determined the column of the matrix, whereas the probed nuclear spin stated determined the row. An example is given in Fig. 7.8c, where the nuclear spin was initialized (vertical line) and probed (horizontal line) in the $|1/2\rangle$ state, giving rise to an element on the diagonal of the matrix. Notice that the diagonal is going from the lower left to the upper right corner. By repeating this procedure several hundred times, we gathered enough data points to plot the 2D matrix (see Fig. 7.8d). Since the relaxation time is much longer than the measurement cycle, most elements are on the diagonal of the matrix. If, however, the MW was in resonance with the $m_I = 3/2 \longleftrightarrow m_I = 1/2$ transition, increased offdiagonal elements will appear, as indicated by blue-green rectangles. Other off-diagonal elements were also observed due to thermal relaxation processes but with much less intensity. Scanning the frequency, in steps of 2 MHz, from 2.3 to 2.5 GHz led to the results presented in Fig. 7.9a. When the microwave frequency hit the resonance of the nuclear qubit transition at 2.45 GHz, we obtained a matrix as shown in Fig. 7.9b, in which off-diagonal elements for the expected transition are clearly observed.

If, however, the microwave power was chosen too large or the pulse width was set too long, the device suffered from heating of the nuclear spin states, resulting in additional off-diagonal elements. In contrary to the resonant condition, the thermal heating affected all four nuclear spin states and can easily be distinguished (see Fig. 7.10).

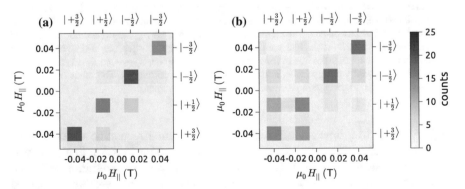

Fig. 7.9 Matrix similar to Fig. 7.8 for 400 sweeps when the microwave frequency was off resonance (**a**) and on resonance (**b**) with the $m_I = 3/2 \longleftrightarrow m_I = 1/2$ transition

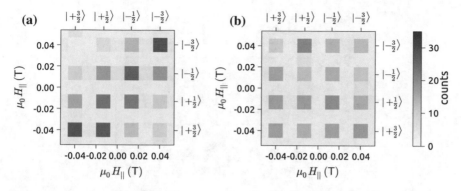

Fig. 7.10 Matrix similar to Fig. 7.8 for 1000 measurement at two different powers, off-resonant to the nuclear qubit transition

7.2.2 Rabi Oscillations

After having found the resonance frequency between the two lowest lying nuclear qubits, we started performing measurements for a fixed frequency as a function of the MW pulse duration τ. To initialize the nuclear spin qubit in its $|+3/2\rangle$ ground state, the external magnetic field is swept back and forth between -75 and 75 mT/s at 100 mT/s (see Fig. 7.11a) until a QTM transition is measured at -38 mT, which is the signature of the $|+3/2\rangle$ qubit state (Fig. 5.7.1a). Using a Rhode & Schwarz SMA100A signal generator, a MW pulse of duration τ is then applied while keeping the external field constant (Fig. 7.11a). The resulting state is detected by sweeping back the external magnetic field in a time scale faster than the measured relaxation times of both nuclear spin states. The sequence was rejected when the final state was not detected due to a missing QTM transition. In order to get a sufficient approximation of the nuclear spin qubit expectation value, the procedure was repeated 100 times for each pulse duration, resulting in coherent Rabi oscillations, as presented in Fig. 7.11b, c for two different microwave powers. The visibility of the measurements presented in Fig. 7.11b, c is $\sim 50\%$.

From Eq. 7.1.9 we see that the Rabi frequency is proportional to the amplitude of the effective magnetic field for zero detuning Δ. Assuming that increasing the microwave power will increase the effective magnetic field, we should observe a monotonic increase of the Rabi frequency with the microwave power. To investigate this behavior we measured the frequency of the Rabi oscillation Ω_R at different injection powers P (see Fig. 7.12). The result shows a linear dependence of Ω_R with $\sqrt{P}$ above 2 mW of injection power. For smaller powers, however, we found a deviation from the this linear curve. One reason could be a nonlinearity in the hyperfine Stark effect or a slight gate voltage drift during the 5 days needed to perform this experiment. Indeed, we will see in the following that the Rabi frequency is extremely sensitive to modifications of the gate voltage because of the Stark effect.

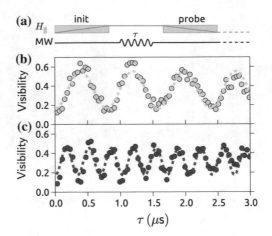

Fig. 7.11 Rabi oscillations of a single nuclear spin qubit. **a** Time dependent external magnetic field $H_{||}$ and pulse sequence generated to observe Rabi oscillations between the two lower states of the nuclear spin qubit having a resonant frequency ν_0. The nuclear spin is first initialized by detecting a conductance jump while sweeping up $H_{||}$ (init sequence). A subsequent MW pulse of frequency ν_0 and duration τ is applied, modifying periodically the hyperfine constant A. It induces an effective oscillating magnetic field resulting in coherent manipulation of the two lower states of the nuclear spin qubit. Finally, $H_{||}$ is swept down to probe the final state of the nuclear spin qubit. **b** Rabi oscillations obtained by repeating the above sequence 800 times for each τ, for two different MW powers, $P_{MW} = 1$ mW and $P_{MW} = 1.58$ mW for the red and violet measurements

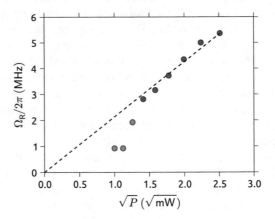

Fig. 7.12 Rabi frequency $\Omega_R/2\pi$ as a function of the microwave power P

7.3 Experimental Discussion of the Hyperfine Stark Effect

7.3.1 DC Gate Voltage Induced Hyperfine Stark Effect

We now present and discuss the study of the visibility of the Rabi oscillations as a function of the applied MW frequency at three different gate voltage values (Fig. 7.13a). As expected from theory (compare Fig. 7.5), the visibility of the Rabi oscillations was largest at the resonant frequency ν_0 and decreases for increasing detuning $\Delta = |\nu - \nu_0|$. However, a clear dependence of the nuclear qubit resonance frequency on the gate voltage is also observed in Fig. 7.13a. This effect can be attributed to the static HF Stark shift, due to the additional electric field induced by the gate voltage, which shows our ability to tune the HF constant A between the electronic spin and the nuclear spin qubit. Notice that only the z-component of the effective magnetic field will modify the level splitting. Applying a gate voltage offset of 10 mV and 16 mV resulted in a shift of $\Delta\nu_0 = 1.72$ and 7.03 MHz respec-

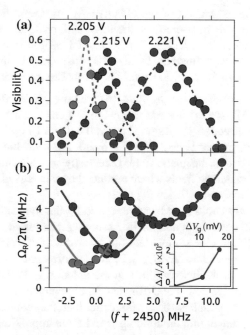

Fig. 7.13 Stark shift of the hyperfine coupling. **a** Rabi oscillations visibility measured at different MW frequencies for three different gate voltages V_g. The resonance shift of the nuclear spin qubit frequency ν_0 is caused by a modification of the hyperfine coupling A due to V_g induced Stark shift. **b** Rabi frequencies Ω_R corresponding to the visibility of (**a**). The continuous lines are fit to the experimental points following the theoretical expression of the Rabi frequency dependence (see main text). The magnitude of the effective magnetic field induced by the oscillating hyperfine constant A due to Stark shift reaches a few hundreds of mT, resulting in Rabi frequencies up to several MHz

tively. Converting this frequency shift into a change of the hyperfine constante gives $\Delta A/A = 5.6 \times 10^{-4}$ for $\Delta V_g = 10$ mV and $\Delta A/A = 2.3 \times 10^{-3}$ for $\Delta V_g = 16$ mV. (see inset Fig. 7.13). Those values can be compared with calculations presented in Sect. 7.4. There we estimate an order of magnitude of $\Delta A/A = 10^{-3}$ for an electric field of 1 mV/nm. The conversion of the back gate voltage into an electric field can be done using the simple formula E = V/d, a gate oxide thickness of 7 nm, and the screening factor of 0.2 (which is a typical value for devices created by electromigration). Doing so, we obtain $\Delta A/A = 2.9 \times 10^{-4}$ for $\Delta V_g = 10$ mV and 4.6×10^{-4} for $\Delta V_g = 16$ mV. Those values are smaller but in the same order of magnitude as the experimental values and therefore within the error bar of the theoretical model. But most importantly, these results show our ability to control the resonance frequency of a single nuclear spin qubit by means of an electric field only.

7.3.2   AC Induced Hyperfine Stark Effect

We turn now to the estimation of the effective AC magnetic field. To do so, the Rabi frequency Ω_R was measured for the three different gate voltage as a function of the detuning Δ (Fig. 7.13b). The horizontal evolution of the minimum of the Rabi oscillations as a function of the MW frequency is induced by the DC Stark shift as explain in Sect. 7.3.1. By further fitting the measurements to the function $\Omega_R/2\pi = \sqrt{(\Delta/2\pi)^2 + (\sqrt{3}g_N\mu_N B_x/h)^2}$, with g_N being the nuclear g-factor (≈ 1.354 for Tb [15]), μ_N the nuclear magneton, we can extract the effective magnetic field in the x-direction B_x. Astonishingly, the data of Fig. 7.13 gives values of $B_x = 62, 98$ and 183 mT for $V_g = 2.205, 2.215$ and 2.221 V, which are up two orders of magnitude higher than magnetic fields created by on-chip micro-coils. In order exclude that those magnetic fields where produced by currents in the vicinity of the spin we were considering the following cases.

(I) The magnetic field could have been generated by the magnetic field component emitted by the microwave antenna itself. Assuming a minimal distance of 10 μm between the antenna and the sample leads to a current of 10 A in order to generate 200 mT using the formula $I = 2\pi r B/\mu_0$. From measurements with a vector network analyzer we know that the insertion loss of the antenna is 35.5 dBm at 2.45 GHz. Considering a microwave power of 0 dBm and an impedance of 50 Ω the current can be estimated to 75 μA, which is 10^5 times smaller than the required current to obtain 200 mT. Moreover, the aluminum bonding wire has an approximate fuse current of 300 mA.

(II) The magnetic field could have been created by the tunnel current through the molecule. This time we can assume a distance of 0.5 nm between the electronic spin and the tunnel current. Using the formula $I = 2\pi r B/\mu_0$ as a rough estimate, results in a required tunnel current of 500 μA. However, the current through the molecule is in the order of 1 nA, which is 5×10^5 times smaller. Even the maximal current through a single molecule, which can be as large as 100 nA, is not sufficient to explain such high magnetic fields.

(III) The magnetic field could have been created by the hyperfine Stark effect, which describes the influence of the electric field on the hyperfine interaction. The hyperfine interaction can be seen as an interaction with an effective magnetic field, which is generated by the electronic spin at the center of the nucleus. Manipulating the interaction constant A by means of an oscillating electric field results in an alternating magnetic field. In order to achieve a magnitude of 200 mT at 0 dBm, the relative variation of the hyperfine constant $\Delta A/A$ should be in the order of the ratio of the corresponding Rabi oscillation to the hyperfine splitting which is 1 MHz/ 2.45 GHz $\approx 10^{-3}$ which would require electric field fluctuations in the order of 1 mV/nm.

The first step towards a verification of the third possibility was to quantify the amplitude of the pulsed oscillating electric field, used to perform the Rabi oscillations. To do so, the full width at half maximum (FWHM) of the dip at the right side of the charge degeneracy point of sample C ($V_g = 2.2$ V in Fig. 5.2.1) was measured as a function of the applied microwave power (see Fig. 7.14a). The observed dip is a signature created by a transition from the inelastic cotunneling between the singlet/triplet state to elastic cotunneling through the singlet state only. In a first approximation, the amplitude of the induced AC voltage is directly proportional to the broadening of the dip. Since the microwave power had to be applied continuously, we could measure only up to a injection power of -20 dBm in order to avoid any damage of the sample. Figure 7.14a shows the evolution of the FWHM from -40 to -20 dBm and an extrapolation up to 0 dBm. From this measurement we see that the induced voltage drop across the molecule is about 2 mV at 0 dBm. Given the size of the molecule to be 1 nm, the generated electric field is estimated to be 2 mV/nm. We will use this value in Eq. 7.4.16 to estimate a relative change of the hyperfine constant to $\Delta A/A = 2 \times 10^{-3}$. This value is in the same order of magnitude than the required value of the consideration given above.

This result emphasized the possibility to use the hyperfine Stark effect to manipulate a single nuclear spin by means of an electric field only. The estimated effective magnetic field in the order of 200 mT and about two orders of magnitude higher

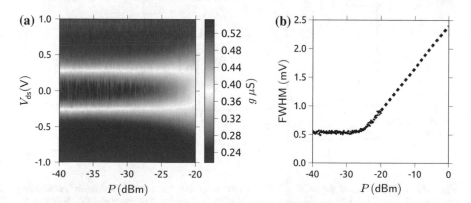

Fig. 7.14 a Conductance through the read-out dot as a function of the source-drain voltage V_{ds} and the applied microwave power P. **b** Evolution of the full width half maximum (FWHM) of the dip in (**a**) as a function of the microwave power P

than the fields generated by on-chip micro-coils, which leads to an increase of the clock-speed of the coherent manipulation.

7.4 Theoretical Discussion of the Hyperfine Stark Effect

The model presented in this section was elaborated in cooperation with Rafik Ballou from the Néel institute and is aimed to give an order of magnitude explanation of the experimental data in Sect. 7.3.2. To keep the derivation as intuitive as possible, rather complicated algebraic calculation were cut out, and only the result will be given.

To determine the magnitude of the hyperfine Stark effect, we used to the following strategy. Starting from the isolated terbium ion, we consider the effect of the ligand field as a perturbation on the electronic configurations. Subsequently, the Stark effect is treated as perturbation on the ligand field ground states. In this way, we derive an expression which connects the mixing of the ground state wavefunctions with the electric field. Afterward, we will evaluate hyperfine interaction with the mixed ground states within first order perturbation theory. Thus, we are able obtain an expression correlating the electric field $\mathcal{E}$ with the change of the hyperfine constant A.

The isolated Tb^{3+} ion possesses a ground state configuration of $4f^8$ and an excited state configuration of $4f^7 5d^1$ (see Fig. 7.15a). The latter arises from an excitation of one 4f electron into the 5d orbital and is about 5.5 eV higher in energy . Moreover, the lowest energy states of each configuration (states having $S = max$ and $L = max$) are split into levels of different J due to the spin-orbit interaction (compare Fig. 3.5.2). To

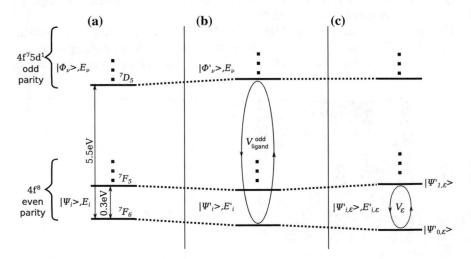

Fig. 7.15 a Illustration of the isolated Tb^{3+} electronic ground state configurations $4f^8$, containing the states Ψ_i), and the first excited configuration $4f^7 5d^1$, composed of the states Φ_ν. **b** Under the influence of a symmetry breaking ligand field $V_{\text{ligand}}^{\text{odd}}$, the ground state and excited state configurations are mixed along with their parities. **c** If an additional electric field, described by the operator $V_{\mathcal{E}}$, is applied, states within ground state configuration are being mixed

distinguish these states from each other, we will use the spectroscopy nomenclature $^{2S+1}X_J$, with $S = \sum_i s_i$, $L = \sum_i l_i$, $J = |L - S|...L + S$, and $X = S, P, D, F$ for $L = 0, 1, 2, 3$. For the Tb^{3+} the two lowest energy multiplets are $^7F_6$ and $^7F_5$, which correspond to states with $S = 3$, $L = 3$, and $J = 6$ or $J = 5$ respectively.

At this point, we want to recall that the parity P of the wavefunction is defined as $P = (-1)^{\sum_i l_i}$ with $l_i = 0, 1, 2, 3, \ldots$ for $s, p, d, f, \ldots$ electrons. Thus, $P = (-1)^{8*3} = 1$ for all the states of the ground configuration $4f^8$ of the Tb^{3+} free ion, whereas $P = (-1)^{7*3+2} = -1$ for all the states of its first excited configuration $4f^7 5d^1$.

If the isolated terbium ion is placed into the electrostatic environment of the molecule, all electronic levels are modified by the ligand field operator V_{ligand}^{odd}. Since the molecule lacks an inversion symmetry, the operator contains contributions of odd parity, which is able to mix the states $|\Psi_i\rangle$ of the ground configuration $4f^8$ with states $|\Phi_\nu\rangle$ of the excited configuration $4f^7 5d^1$ of opposite parity. In first order perturbation theory, the modified ground state multiplets $|\Psi_i'\rangle$ are calculated as:

$$|\Psi_i'\rangle = |\Psi_i\rangle + \sum_\nu \frac{\langle \Phi_\nu | V_{ligand}^{odd} | \Psi_i \rangle}{E_i - E_\nu} |\Phi_\nu\rangle = |\Psi_i\rangle + \sum_\nu \alpha_i^\nu |\Phi_\nu\rangle \quad (7.4.1)$$

where $E_i - E_\nu$ is the energy difference between the states $|\Phi_\nu\rangle$ of the $4f^7 5d^1$ configuration and the state $|\Psi_i\rangle$ of the $4f^8$ ground configuration. Note that without parity breaking, the ligand field operator would have been of even parity and the term $\langle \Phi_\nu | V_{ligand} | \Psi_i \rangle = 0$.

If, furthermore, an external electric field $\mathcal{E}$ is applied, the lowest energy levels of the ground state configuration $4f^8$ (all terms beginning with ^7F) are themselves mixed due to the Stark interaction $V_\mathcal{E} = -d\mathcal{E}$. In first order of perturbation, the in this way altered wavefunctions $|\Psi_{i\mathcal{E}}'\rangle$ are determined as:

$$|\Psi_{i\mathcal{E}}'\rangle = |\Psi_i'\rangle + \sum_j \frac{\langle \Psi_j' | V_\mathcal{E} | \Psi_i' \rangle}{E_i' - E_j'} |\Psi_j'\rangle = |\Psi_i'\rangle + \sum_j \beta_i^j |\Psi_j'\rangle$$

$$= |\Psi_i\rangle + \sum_\nu \alpha_i^\nu |\Phi_\nu\rangle + \sum_j \beta_i^j |\Psi_j\rangle + \sum_j \beta_i^j \sum_\nu \alpha_j^\nu |\Phi_\nu\rangle \quad (7.4.2)$$

At this point we have successfully established the correlation of the electric field $\mathcal{E}$ with the mixing of the ground state wavefunctions. All what remains is evaluation of the hyperfine splitting using the perturbed ground states $|\Psi_{i\mathcal{E}}'\rangle$. To do so, we have to determine the expression of the hyperfine Hamiltonian first. Generally, the hyperfine interaction can be considered as a change of the potential energy of the nuclear magnetic moment μ_1, exposed to the magnetic field B_{elec}, which is created by the ensemble of the electrons in the 4f shell. Therefore, the hyperfine Hamiltonian can be written as:

$$H_{hf} = -\mu_1 B_{elec} \quad (7.4.3)$$

The magnetic field operator $\boldsymbol{B}_{\text{elec}}$ consists of two independent contributions, an orbital contribution $\boldsymbol{B}_{\text{orbit}}$ coming from the motion of the electrons around the core, and a spin contribution $\boldsymbol{B}_{\text{spin}}$ resulting from the magnetic dipole field of the electron's spin. Since the probability density of 4f electrons is zero at the core, there is no contact interaction. To cut down the problem, we are going to consider the orbital contribution first. The magnetic field $\boldsymbol{B}_i$, created by a moving electron i at velocity $\boldsymbol{v}_i$ and distance $\boldsymbol{r}_i$ of the atomic core, is given by the law of Biot-Savart:

$$B_i = \frac{\mu_0}{4\pi} e\boldsymbol{v}_i \times \frac{\boldsymbol{r}_i}{r_i^3} \tag{7.4.4}$$

Since $e\boldsymbol{v}_i \times \boldsymbol{r}_i = -2\frac{e}{2m}\boldsymbol{r}_i \times m\boldsymbol{v}_i = -2\mu_B\boldsymbol{l}_i$ and thus the orbital contribution becomes:

$$B_{\text{orbit}} = -\frac{\mu_0}{4\pi} 2\mu_B \sum_i \frac{\boldsymbol{l}_i}{r_i^3} \tag{7.4.5}$$

Now we turn to the spin contribution $\boldsymbol{B}_{\text{spin}}$. We assume that the spin is localized on each electron, so that the magnetic field seen by the nucleus is just the sum of the magnetic field created by each magnetic moment $\boldsymbol{\mu}_s^i$ at the distance r_i.

$$B_{\text{spin}} = -\frac{\mu_0}{4\pi} \sum_i \frac{\boldsymbol{\mu}_s^i}{r_i^3} - \frac{3\boldsymbol{r}_i(\boldsymbol{\mu}_s^i\boldsymbol{r}_i)}{r_i^5} \tag{7.4.6}$$

Substituting $\boldsymbol{\mu}_s^i = -2\mu_B\boldsymbol{s}_i$ and $\boldsymbol{\mu}_I = g_N\mu_N\boldsymbol{I}$, we obtain following hyperfine Hamiltonian

$$H_{\text{hf}} = -\boldsymbol{\mu}_I\left(\boldsymbol{B}_{\text{orbit}} + \boldsymbol{B}_{\text{spin}}\right) \tag{7.4.7}$$

$$= a\sum_i (N_i/r_i^3) \cdot \boldsymbol{I} \tag{7.4.8}$$

where $a = \frac{\mu_0}{4\pi} 2g_N\mu_N\mu_B$ is a constant, $\boldsymbol{I}$ is the nuclear spin and $N_i = \boldsymbol{l}_i - \boldsymbol{s}_i + 3\boldsymbol{r}_i(\boldsymbol{s}_i \cdot \boldsymbol{r}_i)/r_i^2$ is the operator accounting for the interaction with the ith electron having the spin $\boldsymbol{s}_i$ and the angular momentum $\boldsymbol{l}_i$ at a distance $\boldsymbol{r}_i$. On the quantum states from which the electronic degrees of freedom can be factored out into a state $|\Psi_0\rangle$, the electronic part of the hyperfine interaction is given as $\langle\Psi_0|N|\Psi_0\rangle\langle 1/r^3\rangle$, where $N = \sum_i N_i$ and where the radial integral $\langle 1/r^3\rangle$ is a constant within the same electronic configuration. In the absence of any parity breaking interaction the electronic state $|\Psi_0\rangle$ has a well defined parity P. We recall that $P = (-1)^{\sum_i l_i}$ with $l_i = 0, 1, 2, 3, \ldots$ for $s, p, d, f, \ldots$ electrons. Thus, $P = 1$ for all the states of the ground configuration $4f^8$ of the Tb^{3+} free ion, whereas $P = -1$ for all the states of its first excited configuration $4f^7 5d^1$. It is also crucial to recall that the matrix elements of an operator O of even (resp. odd) parity, i.e. invariant (resp. reversed) under the space inversion, are non zero solely between states with the same (resp. opposite) parity. The position vector $\boldsymbol{r}$ is reversed by space inversion whereas the orbital $\boldsymbol{l}$

and spin s moment operators are invariant, which implies that the dipole electric moment operator d is of odd parity but that the operator N is of even parity. It is a matter of standard use of the Racah algebra [16] to compute the matrix element of the spherical components N_q ($q = -1, 0, 1$) of the operator N between any two states of an electron shell. Within the Russel-Saunders coupling scheme and by making use of the Wigner-Eckart theorem one computes

$$\langle 4f^8 \xi SLJM | N_q | 4f^8 \xi' S'L'J'M' \rangle = (-1)^{J-M} \begin{pmatrix} J & 1 & J' \\ -M & q & M' \end{pmatrix} \times$$

$$\times (4f^8 \xi SLJ \| L - (10)^{\frac{1}{2}} \sum_i (s^{(1)} C^{(2)})_i^{(1)} \| 4f^8 \xi' S'L'J')$$

(7.4.9)

with

$$(\cdots \| L \| \cdots) = \delta(\xi, \xi') \delta(S, S') \delta(L, L') \times$$

$$\times (-1)^{S+L+J+1} ([J][J'])^{\frac{1}{2}} (L(L+1)(2L+1))^{\frac{1}{2}} \begin{Bmatrix} S & L & J \\ 1 & J' & L' \end{Bmatrix}$$

and

$$\left(\cdots \| \sum_i (s^{(1)} C^{(2)})_i^{(1)} \| \cdots\right) = ([J][1][J'])^{\frac{1}{2}} ([1][2])^{-\frac{1}{2}} \begin{Bmatrix} S & S' & 1 \\ L & L' & 2 \\ J & J' & 1 \end{Bmatrix} (s\|s\|s)(l\|C^{(2)}\|l) \times$$

$$\times (4f^8 \xi SL \| W^{(12)} \| 4f^8 \xi' S'L')$$

where $[x] = 2x + 1$, $\delta(X, X') = 1$ if and only if $X = X'$ and $= 0$ otherwise, $(:::)$, $\{:::\}$ and $\{\vdots\vdots\vdots\}$ stand for the 3j, 6j and 9j symbols, and the reduced matrix elements of the tensor operator $W^{(12)}$ are tabulated [17] or can be computed by making use of the coefficients of fractional parentages [16].

We shall now consider that the electronic wavefunction is exposed to the ligand field and an external electric field E resulting in the Stark interaction $V_E = -d \cdot E$. Since the molecule lacks an inversion symmetry the electrostatic interactions with the ligand field contains contributions of odd parity $V_{\text{ligand}}^{\text{odd}}$, which mixes the states $|\Psi_i\rangle$ of the ground configuration $4f^8$ with states $|\Phi_v\rangle$ of the excited configuration $4f^7 5d^1$ of opposite parity. In first order perturbation theory the new wavefunction $|\Psi_i'\rangle$ is approximated as

$$|\Psi_i'\rangle = |\Psi_i\rangle + \sum_v \frac{\langle \Phi_v | V_{\text{ligand}}^{\text{odd}} | \Psi_i \rangle}{E_i - E_v} |\Phi_v\rangle = |\Psi_i\rangle + \sum_v \alpha_i^v |\Phi_v\rangle, \qquad (7.4.10)$$

where $E_i - E_v$ is the energy difference between the states $|\Phi_v\rangle$ of the $4f^7 5d^1$ configuration and the state $|\Psi_i\rangle$ of the $4f^8$ ground configuration. Owing to this admixture, the states of the ground configuration $4f^8$ are themselves mixed under an applied electric field as

$$|\Psi'_{iE}\rangle = |\Psi'_i\rangle + \sum_j \frac{\langle\Psi'_j|V_E|\Psi'_i\rangle}{E'_i - E'_j}|\Psi'_j\rangle = |\Psi'_i\rangle + \sum_j \beta_i^j|\Psi'_j\rangle$$

$$= |\Psi_i\rangle + \sum_\nu \alpha_i^\nu|\Phi_\nu\rangle + \sum_j \beta_i^j|\Psi_j\rangle + \sum_j \beta_i^j \sum_\nu \alpha_j^\nu|\Phi_\nu\rangle, \quad (7.4.11)$$

now to first order in perturbation in V_E with respect to $V_{\text{ligand}}^{\text{odd}}$. The influence of the Stark effect on the hyperfine coupling can be evaluated by calculating the matrix element of the operator N on the perturbed state $|\Psi'_{0E}\rangle = |\Psi'_0\rangle + \sum_j \beta_0^j|\Psi'_j\rangle = |\Psi_0\rangle + \sum_\nu \alpha_0^\nu|\Phi_\nu\rangle + \sum_j \beta_0^j|\Psi_j\rangle + \sum_j \beta_0^j \sum_\nu \alpha_j^\nu|\Phi_\nu\rangle$:

$$\langle\Psi'_{0E}|N|\Psi'_{0E}\rangle = \langle\Psi_0|N|\Psi_0\rangle + \sum_{j\neq0}(\beta_0^j\langle\Psi_0|N|\Psi_j\rangle + \beta_0^{j\star}\langle\Psi_j|N|\Psi_0\rangle) + \cdots,$$
$$(7.4.12)$$

where contributions involving products of the coefficients α_i^ν and β_i^j are ignored as being negligible. It is emphasized that $\sum_\nu \alpha_0^\nu\langle\Psi_0|\sum_i(N_i/r_i^3)|\Phi_\nu\rangle+$ complex conjugate $=0$, because $|\Psi_0\rangle$ and $|\Phi_\nu\rangle$ are of opposite parity and $\sum_i(N_i/r_i^3)$ is of even parity. Assuming that $E'_0 - E'_j \approx E_0 - E_j$ and $E_0 - E_\nu \approx \Delta E_{4f^8\to4f^75d^1}$ then using the closure relation $\sum_\nu |\Phi_\nu\rangle\langle\Phi_\nu| = 1$, the coefficient β_0^j can be approximated as

$$\beta_0^j = \frac{\langle\Psi'_j|V_E|\Psi'_0\rangle}{E'_0 - E'_j} \quad (7.4.13)$$

$$= \frac{\{\langle\Psi_j| + \sum_\tau\langle\Phi_\tau|\frac{\langle\Psi_j|V_{\text{ligand}}^{\text{odd}}|\Phi_\tau\rangle}{E_0-E_\tau}\}V_E\{|\Psi_0\rangle + \sum_\nu \frac{\langle\Phi_\nu|V_{\text{ligand}}^{\text{odd}}|\Psi_0\rangle}{E_0-E_\nu}|\Phi_\nu\rangle\}}{E'_0 - E'_j}$$

$$\approx 2\frac{\langle\Psi_j|V_E V_{\text{ligand}}^{\text{odd}}|\Psi_0\rangle}{(E_0 - E_j)\Delta E_{4f^8\to4f^75d^1}},$$

The change in the hyperfine interaction may finally be written as

$$\langle\Psi'_{0E}|A\boldsymbol{J}\cdot\boldsymbol{I}|\Psi'_{0E}\rangle = (1 + \Delta A/A)\langle\Psi_0|A\boldsymbol{J}\cdot\boldsymbol{I}|\Psi_0\rangle \quad (7.4.14)$$

with

$$\Delta A/A \approx 4\sum_j \frac{\langle\Psi_j|V_E V_{\text{ligand}}^{\text{odd}}|\Psi_0\rangle}{(E_0 - E_j)\Delta E_{4f^8\to4f^75d^1}}\frac{\langle\Psi_0|N|\Psi_j\rangle}{\langle\Psi_0|N|\Psi_0\rangle} \quad (7.4.15)$$

In general the crystal field experienced by the excited configuration $4f^75d^1$ is about ten times larger [18] than the one experienced by the electrons of the ground configuration $4f^8$. It is then reasonable to expect that the effect of $V_{\text{ligand}}^{\text{odd}}$ amounts to around $1-2$ eV in energy. On the other hand, given the size of the electronic orbits, which is within the range 0.1–0.2 nm, and the expression of the dipole operator $d = -er$, the strength of V_E under an electric field E measured in mV/nm is estimated in eV to $(1 - 2) \cdot 10^{-4}$ E. The excited configuration $(4f^7 5d^1)$ is separated from the ground configuration $(4f^8)$ by about $\Delta E_{4f^8 \to 4f^7 5d^1} = 5.5$ eV. The quantity $4\langle \Psi_j | V_E V_{\text{ligand}}^{\text{odd}} | \Psi_0 \rangle / \Delta E_{4f^8 \to 4f^7 5d^1}$ thus is estimated to $(1.8 \pm 1.1) 10^{-4}$ (eV) E with E given in mV/nm. If furthermore, we consider only the states of the ground multiplet $^7F_6$ and those of the first excited $^7F_5$ multiplet then only two excited states are mixed by the electric field with the ground state, with $E_0 - E_{j=1} \approx -0.06$ eV and $\langle \Psi_0 | N | \Psi_{j=1} \rangle / \langle \Psi_0 | N | \Psi_0 \rangle = -1/\sqrt{6}$ for the first and $E_0 - E_{j=2} \approx -0.3$ eV and, making use of the Eq. 7.4.9, $\langle \Psi_0 | N | \Psi_{j=2} \rangle / \langle \Psi_0 | N | \Psi_0 \rangle = -0.41576$ for the second. With all these numbers we may reasonably expect a change in the hyperfine constant in the order of

$$\frac{\Delta A}{A} \approx 10^{-3} \, E(\text{mV/nm}) \tag{7.4.16}$$

The result is in the same order of magnitude than the experimental value and shows that the observed nuclear spin response to the electric field is explainable by the hyperfine Stark effect.

7.5 Dephasing Time T_2^*

7.5.1 Introduction

In this section, we are going to present measurements of the dephasing time T_2^* of the nuclear spin qubit. The dephasing time is equal to the duration over which the time average coherence of the quantum superposition is preserved. But before turning to the discussion of the experimental results, a brief review about the dephasing of an effective spin 1/2 and the experimental access to this quantity is given. To do so, we will follow the common approach by starting from the time evolution of the 2×2 density matrix ρ:

$$i\hbar \frac{d\rho}{dt} = [H, \rho] = H\rho - \rho H = \frac{\hbar}{2} \left[\begin{pmatrix} \Delta & \Omega \\ \Omega e^{-i\phi} & -\Delta \end{pmatrix} \rho - \rho \begin{pmatrix} \Delta & \Omega \\ \Omega e^{-i\phi} & -\Delta \end{pmatrix} \right] \tag{7.5.1}$$

where H is the Hamiltonian described in Eq. 7.1.8. Expanding this matrix equation and substituting

$$\langle \sigma_x \rangle = (\rho_{21} + \rho_{12}) \tag{7.5.2}$$

$$\langle \sigma_y \rangle = i \, (\rho_{21} - \rho_{12}) \tag{7.5.3}$$

$$\langle \sigma_z \rangle = \rho_{22} - \rho_{11} \tag{7.5.4}$$

we get the equations of motion in the rotation frame:

$$\langle \sigma_x \rangle = \Delta \langle \sigma_y \rangle \tag{7.5.5}$$

$$\langle \sigma_y \rangle = -\Delta \langle \sigma_x \rangle + \Omega \langle \sigma_z \rangle \tag{7.5.6}$$

$$\langle \sigma_z \rangle = -\Omega \langle \sigma_y \rangle \tag{7.5.7}$$

These equations describe the motion of the spin exposed to an alternating field, however, the effects of relaxation and decoherence are still missing. In 1946, Felix Bloch extended this set of equations by empirical terms to allow for the relaxation to equilibrium. He assumed that the relaxations along the z-axis and in the $x - y$ plane happen at different rates, which are designated as $1/T_1$ and $1/T_2$ for the z-axis and the $x - y$ plane respectively. Including these terms results in the Bloch equations:

$$\langle \sigma_x \rangle = -\Delta \langle \sigma_y \rangle - \frac{\langle \sigma_x \rangle}{T_2} \tag{7.5.8}$$

$$\langle \sigma_y \rangle = \Delta \langle \sigma_x \rangle + \Omega \langle \sigma_z \rangle - \frac{\langle \sigma_y \rangle}{T_2} \tag{7.5.9}$$

$$\langle \sigma_z \rangle = -\Omega \langle \sigma_y \rangle - \frac{\langle \sigma_z \rangle}{T_1} \tag{7.5.10}$$

In case of no alternating field $\Omega = 0$ one can show that the solution to these equations is:

$$\langle \sigma_x \rangle = \langle \sigma_x \rangle_{t=0} \, cos(\Delta t) e^{-t/T_2} \tag{7.5.11}$$

$$\langle \sigma_y \rangle = \langle \sigma_y \rangle_{t=0} \, sin(\Delta t) e^{-t/T_2} \tag{7.5.12}$$

$$\langle \sigma_z \rangle = \langle \sigma_z \rangle_{t=0} \, (1 - e^{-t/T_1}) \tag{7.5.13}$$

Equations 7.5.11–7.5.13 describe the precession of a spin 1/2 with the detuning Δ around the z-axis. This precession is damped at a rate $1/T_2$ in the $x - y$-plane and with the rate $1/T_1$ along the z-axis. To measure the relaxation in the $x - y$-plane (free induction decay) a series of operations is performed. First, the spin is prepared along the $+z$-axis in the Bloch sphere at $t = 0$. Subsequently, we turn the spin into the equatorial plane using a MW pulse and thus create a superposition between the two spin states. The duration of this pulse was adjusted to perform a 90° rotation around the x-axis, which is why this type of pulse is referred to as a $(\pi/2)$ pulse (Fig. 7.16a). Afterward, we are waiting for the time τ, leading to the precession of the spin according to Eqs. 7.5.11–7.5.12 around the z-axis at the frequency Δ. Notice that $\langle \sigma_z \rangle$ remains zero and only $\langle \sigma_x \rangle$ and $\langle \sigma_y \rangle$ are changing. Then a second $\pi/2$ pulse is rotating the spin back onto the z-axis. This operation transforms the former value of $\langle \sigma_y \rangle$ to $\langle \sigma_z \rangle$, which is measured subsequently. Repeating this pulse sequence (Fig. 7.16b) for different values of τ and measuring the resulting value of $\langle \sigma_z \rangle$ leads to oscillations with a period of $1/\Delta$ — the so called Ramsey fringes (Fig. 7.16c). In the case of a single spin, many measurements are averaged to obtain the expectation value $\langle \sigma_z \rangle$. Due to the changing environmental influence in each measurement, the spin performs rotations with slightly different angles at a given time τ between the two $\pi/2$ pulses. This dephasing mechanism between subsequent

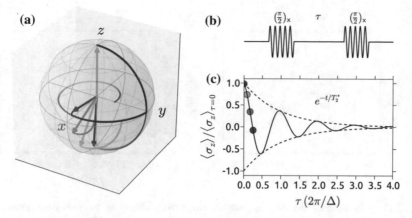

Fig. 7.16 **a** Bloch sphere trajectory of the spin wavefunction during the experiment. The Ramsey fringes are measured by applying a sequence of two MW pulses (**b**) to a spin, which was initially oriented along the z-axis. The first MW pulse rotates the vector by $90°$ into the $x − y$-plane, which is equivalent to a superposition of the two spin states. Waiting for a time τ, causes a damped precession of the Bloch vector with the frequency $2\pi/\Delta$ and at the rate T_2^*. The second MW pulse rotates the spin again by 90° around the x-axis, thus, mapping $\langle\sigma_y\rangle$ on $\langle\sigma_z\rangle$. **c** Repeating this sequence for different τ and measuring the resulting expectation value $\langle\sigma_z\rangle$ leads to oscillations decaying with e^{-t/T_2^*}

measurements leads to a decay faster than the decoherence time. The envelope of the oscillation is modeled by the function $exp(-t/T_2^*)$, where T_2^* is the dephasing time.

7.5.2 Experimental Results

From the previous section we know that the oscillation frequency of the Ramsey fringes is equal to the detuning $\Delta/2\pi$. Therefore, in order to adjust the oscillation period, the precise position of the resonance frequency v_0 had first to be obtained. This was done by measuring the visibility of the Rabi oscillations as function of the frequency at a microwave power of 0 dBm (see Fig. 7.17a). By fitting a Lorentzian to the obtained data points, we found the maximum at 2449 MHz for $V_g = 2.205$ V. Afterward, we detuned the microwave source by 100 kHz in order to see Ramsey fringes with an oscillation period of 10 μs. In the next step we measured a full Rabi oscillation at $v = 2448.9$ MHz to determine the duration of the $\pi/2$ pulse (see Fig. 7.17b). By fitting the Rabi oscillation to a sine function, the duration of the $\pi/2$ pulse can be obtained and was $\simeq$284 ns.

Having calibrated the $\pi/2$ pulse at the microwave frequency of 2448.9 MHz, we measured the Ramsey fringes following the sequence presented in Fig. 7.18a. First the nuclear spin qubit was initialized by sweeping the magnetic field back and forth until the nuclear spin was in the $|+3/2\rangle$ state. Subsequently, two $\pi/2$ MW pulses were generated with the inter-pulse delay τ. At last, the final state was probed sweeping the magnetic field back to its initial value, while checking for a QTM transition.

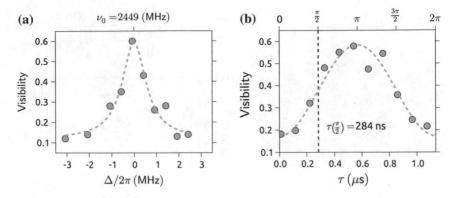

Fig. 7.17 **a** To calibrate the detuning, we measured the visibility of the Rabi oscillations as a function of the frequency at 0 dBm microwave power. The maximum of the visibility corresponds to zero detuning and was found at 2449 MHz. **b** In order to obtain Rabi oscillations with a period of 10 μs, we detuned the microwave source by 100 kHz to 2448.9 MHz and recorded a full period of a Rabi oscillation. Fitting the data to a sine function gave rise to a $\pi/2$ pulse length of 284 ns

If no QTM event was observed, the measurement was rejected. To obtain a good approximation of the expectation value this procedure was repeated 100 times for each inter-pulse delay τ, resulting in the Ramsey fringes as shown in Fig. 7.18b. The measurements exhibit an exponentially decaying cosine function. By fitting the data to $y = cos((\Delta/2\pi)t)exp(-t/T_2^*)$, we extracted a dephasing time $T_2^* \approx 64$ μs.

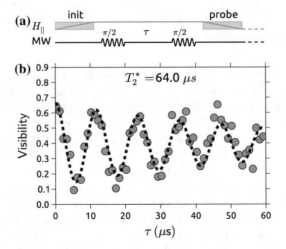

Fig. 7.18 **a** Time dependent external magnetic field $H_\parallel$ and pulse sequence generated to measure the Ramsey fringes. Initialization and probe of the nuclear spin qubit are performed using the identical protocol explained in Fig. 3a. The MW sequence consists of two $\pi/2$ pulses, with an increasing inter-pulse delay τ. **b** Ramsey interference fringes obtained by repeating the procedure of (**a**) 800 times. $V_g = 2.205$ V, corresponding to a Rabi frequency $\Omega_R = 1.136$ MHz and a resonant frequency $\nu_0 = 2.449$ MHz of the nuclear spin qubit. The measured coherence time $T_2^* \approx 64$ μs

Detailed studies suggest that the mayor contribution to the dephasing was caused by charge noise of the oxide and bit noise of the digital to analog converter at the gate terminal. The amplitude of the latter is estimated to be ± 1 bit resulting in gate voltage fluctuations of $\Delta V_g(\text{bit noise}) = \pm 153\ \mu\text{V} \rightarrow$ Fig. 7.13 $\rightarrow \pm 26.2$ kHz $\rightarrow g_N \mu_N B_{\text{eff}}/h \rightarrow \pm 2.6$ mT. Now we turn to the estimation of the noise generated by charges trapped in the gate oxide. If charges are trapped far away from the molecule those fluctuations are small, their frequency, however, is larger due to the multitude of available trapping sites. From our measurements we extracted that within the time scale of averaging over 1 data point we observed an effective gate voltage fluctuation of $\Delta V_g(\text{bit noise}) = \pm 500\ \mu\text{V} \rightarrow$ Fig. 7.13 $\rightarrow \pm 85.9$ kHz $\rightarrow g_N \mu_N B_{\text{eff}}/h \rightarrow \pm 8.6$ mT. Moreover, a charge can be trapped in the close vicinity of the molecule, leading to a gate voltage shift so large that nuclear spin is completely shifted out of resonance. Since the available sites in the close vicinity of the molecular are very few, this event happens in average every 1 to 2 days. Those events will not necessarily increase the decoherence since the changes are so drastic that we recalibrate the resonance frequency every time they occurred. However, they make the measurement of a complete series of Rabi and Ramsey oscillations, which took about 4 days, extremely difficult and time consuming.

In future devices we wil make use of more stabe gate oxides and well stabilized DA converters, which should increase the dephasing time by at least 2 orders of magnitude.

7.5.3 Outlook

In order to enhance the dephasing time T_2^*, the coupling to the environment must be attenuated. This can be achieve by actively controlling the time evolution of the spin precession. This so called dynamical decoupling relies on a series of MW pulses, which is periodically turning the spin [19]. One can show that environmental interactions, which happen on a time scale longer than the pulse series, are canceled out. The most common of these pulse sequences to dynamically decouple the spin from the environment was presented by Hahn [20] and involves as series of three pulses as shown in Fig. 7.19b. To visualize the experiment, we make again use of the Bloch sphere representation (see Fig. 7.19a). The first step, prior to the pulse sequence, is the initialization of the spin. In this example we will use the vector pointing along the $+z$ direction as initial state (grey vector in Fig. 7.19b). By applying a $\pi/2$ pulse, the Bloch vector will be rotated by $90°$ around the x-axis, thus creating a linear superposition state (blue curve). This state is left to a free evolution during the time τ and decoheres at a rate of T_2. After a waiting time τ, the vector has rotated by an angle ϕ in the equatorial plane, which is different every time we perform the experiment due to the fluctuations of the magnetic field along z. The second MW pulse (red) rotates the vector by an angle of $180°$ around x. This operation compensates any difference in ϕ, since vectors which were delayed, due to a slightly smaller magnetic field in z direction (dark red arrow), are now in advance. Followed

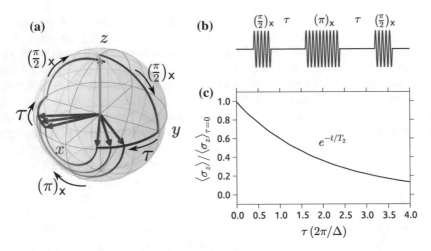

Fig. 7.19 a Trajectory of the Bloch vector in the Hahn spin echo experiment, in which a sequence of three MW pulses (**b**) is applied to a spin, initially aligned along the $+z$-axis. First, the vector is rotated by $90°$ into the equatorial plane using a $\pi/2$ pulse, thus, creating a linear superposition of the two spin states. Afterward, the vector performs a free precession around z for an interval τ, while being damped at the rate e^{-t/T_2}. During this time, magnetic field fluctuations along z result in slightly different precession angles from one measurement to another. A second MW pulse rotates the spin by an angle $180°$ around x, hence, ending up again in the x-y plane. Due to this operation, vectors which were formerly retarded to to a slightly smaller magnetic field are now in advance. Thus, after waiting for the same period τ, the vector will be aligned along the $-y$-axis. Finally, we project it back to the z-axis using a second $\pi/2$ pulse. **c** Repeating this sequence for different τ leads to an exponentially decaying spin echo signal of e^{-t/T_2}

by a second free precession of duration τ, the Bloch vector will arrive at $-y$ no matter what the local magnetic field was, as long as it remained constant on the time scale of the pulse series. Hence, all magnetic field fluctuations, which were much slower than the pulse sequence, are eliminated. Finally, the vector is rotated by $90°$ around x, which brings it back to its original position. Yet, the size of the vector is reduced due to decoherence within the $x - y$ plane, resulting in an exponentially decaying spin echo signal (see Fig. 7.19c). However, the characteristic time of the decay is the decoherence time T_2, which is much longer than T_2^* and can theoretically be extended to its fundamental limit $T_2 \leq 2T_1$.

The measurement protocol will be similar to the Ramsey experiment, but with an altered pulse sequence. We expect to eliminate the rather slow gate voltage fluctuations, which were transformed into magnetic field fluctuations by the hyperfine Stark effect and, hence, we should observe a T_2, which is much larger than T_2^*. However, the experimental realization of this experiment has not been performed yet, and could not be presented in my manuscript.

7.6 Summary

In this chapter, we presented the first quantum manipulation of a single nuclear spin qubit in a single-molecular magnet. To overcome the technical problem of generating high magnetic field amplitudes, we proposed and demonstrated the possibility to use the Stark shift of the hyperfine coupling to not only tune the level splitting of our nuclear spin qubit, but also to generate a large effective AC magnetic field at the nucleus. Using local AC electric fields, we performed electrical quantum manipulations of a single nuclear spin qubit at MHz frequencies with a coherence time $T_2^* \simeq 64$ μs. These results open the way to a fast coherent manipulation of a nuclear spin qubit as well as the opportunity to control the entanglement between different single nuclear spin qubits by tuning their resonance frequency using AC and DC gate voltages, by means of the Stark shift of the hyperfine coupling. Since this was only possible due to the unique electrostatic environment of a single molecule magnet, these results will hopefully make molecular based qubits serious candidates for quantum information processing.

References

1. T. Obata, M. Pioro-Ladrière, T. Kubo, K. Yoshida, Y. Tokura, S. Tarucha, Microwave band on-chip coil technique for single electron spin resonance in a quantum dot. Rev. Sci. Instrum. **78**, 104704 (2007)
2. J.J. Pla, K.Y. Tan, J.P. Dehollain, W.H. Lim, J.J.L. Morton, F.A. Zwanenburg, D.N. Jamieson, A.S. Dzurak, A. Morello, High-fidelity readout and control of a nuclear spin qubit in silicon. Nature **496**, 334–338 (2013)
3. K.C. Nowack, F.H.L. Koppens, Y.V. Nazarov, L.M.K. Vandersypen, Coherent control of a single electron spin with electric fields. Science (New York, N.Y.) **318**, 1430–1433 (2007)
4. L. Meier, G. Salis, I. Shorubalko, E. Gini, S. Schön, K. Ensslin, Measurement of Rashba and Dresselhaus spinâĂŞorbit magnetic fields. Nat. Phys. **3**, 650–654 (2007)
5. Y. Kato, R.C. Myers, D.C. Driscoll, A.C. Gossard, J. Levy, D.D. Awschalom, Gigahertz electron spin manipulation using voltage-controlled g-tensor modulation. Science (New York, N.Y.) **299**, 1201–1204 (2003)
6. E. Laird, C. Barthel, E. Rashba, C. Marcus, M. Hanson, A. Gossard, Hyperfine-mediated gate-driven electron spin resonance. Phys. Rev. Lett. **99**, 246601 (2007)
7. C.P. Slichter, *Principles of Magnetic Resonance* (Springer, Berlin, 1978)
8. J. Johansson, P. Nation, F. Nori, QuTiP: An open-source Python framework for the dynamics of open quantum systems. Comput. Phys. Commun. **183**, 1760–1772 (2012)
9. J.R. Johansson, P.D. Nation, F. Nori, QuTiP 2: A Python framework for the dynamics of open quantum systems. Comput. Phys. Commun. **184**, 1234 (2013)
10. R. Haun, J. Zacharias, Stark effect on cesium-133 hyperfine structure. Phys. Rev. **107**, 107–109 (1957)
11. B.E. Kane, A silicon-based nuclear spin quantum computer. Nature **393**, 133–137 (1998)
12. R. Rahman, C. Wellard, F. Bradbury, M. Prada, J. Cole, G. Klimeck, L. Hollenberg, High precision quantum control of single donor spins in silicon. Phys. Rev. Lett. **99**, 036403 (2007)
13. N. Ishikawa, M. Sugita, W. Wernsdorfer, Quantum tunneling of magnetization in lanthanide single-molecule magnets: bis(phthalocyaninato)terbium and bis(phthalocyaninato)dysprosium anions. Angew. Chem. (International ed. in English) **44**, 2931–2935 (2005)

14. C. Hutchison, E. Wong, Paramagnetic resonance in rare earth trichlorides. J. Chem. Phys. **29**, 754 (1958)
15. J.M. Baker, J.R. Chadwick, G. Garton, J.P. Hurrell, E.p.r. and Endor of TbFormula in thoria. Proc. R. Soc. A Math. Phys. Eng. Sci. **286**, 352–365 (1965)
16. B. Judd, *Operator Techniques in Atomic Spectroscopy* (McGraw-Hill Book Compny, Inc., New York, 1963)
17. J.A. Tuszynski, *Spherical Tensor Operators: Tables of Matrix Elements and Symmetries* (World Scientific, Singapore, 1990). ISBN 9810202830
18. R.M. Macfarlane, Optical Stark spectroscopy of solids. J. Lumin. **125**, 156–174 (2007)
19. L. Viola, S. Lloyd, Dynamical suppression of decoherence in two-state quantum systems. Phys. Rev. A **58**, 2733–2744 (1998)
20. E. Hahn, Spin echoes. Phys. Rev. **80**, 580–594 (1950)

Chapter 8
Conclusion and Outlook

In this thesis I developed an entire experimental setup to measure and manipulate quantum properties of single molecule magnets. Equipped with a new dilution refrigerator to meet the needs of ultra sensitive mesoscopic experiments, I designed and constructed innovative equipments such as miniaturized three dimensional vector magnets, capable of generating static magnetic fields in the order of a Tesla and allowing for sweep-rates larger than one Tesla per second.[1] Furthermore, I optimized the noise-filtering system and the signal amplifiers in order to suppress as much as possible the electronic noise pick-up.

After a thorough testing and approving of every part in the measurement chain, I turned to the fabrication of a single-molecule magnet spin-transistor, being probably one of the smallest devices presented in the field of organic spintronics. However, its tiny size of only 1 nm and the technological limitations of state of the art nano-fabrication made it extremely challenging to build such a device. Despite this difficult conditions, I was able to perform experiments on three different samples, which demonstrated the feasibility and reproducibility of those cutting edge devices, even though the yield was rather small.

The extreme sensitivity of the molecular spin transistor, opened a path to access pristine quantum properties of an isolated single-molecule magnet, such as read-out of its quantized magnetic moment and the detection of the quantum tunneling of magnetization.

But most important, we were able to detect the four different quantum states of an isolated ^{159}Tb nuclear spin. By reading out the nuclear spin states much faster than the relaxation time T_1, we were able to measure the nuclear spin trajectory, revealing quantum jumps between the four different nuclear qubit states at a timescale of seconds. Finally, a post-treatment and statistical averaging of this data yielded the

---

[1]The maximum sweep-rate was tested in liquid helium.

© Springer International Publishing Switzerland 2016
S. Thiele, *Read-Out and Coherent Manipulation of an Isolated Nuclear Spin*,
Springer Theses, DOI 10.1007/978-3-319-24058-9_8

relaxation times T_1 of a few tens of seconds, which were resolved for each nuclear spin state individually.

However, the underlying physics, leading to the relaxation of the nuclear spin, remained hidden in the statistical average. In order to extract this very information, we developed a quantum Monte-Carlo code taking the specific experimental conditions into account. By fitting the statistical average of the simulations to the experimental data, we could deduce that the mechanism, dominating the relaxation process, was established over a coupling between the nuclear spin to the electrons tunneling through the read-out quantum dot. Using this knowledge, we could demonstrate that the experimental relaxation times could be modified just by changing the amount of tunnel electrons per unit time.

After having thoroughly investigated the quantum properties of an isolated nuclear spin using a passive read-out only, we wanted to take our experiments to the next level by actively manipulating the nuclear spin in a coherent manner. To overcome the technical problem of generating high magnetic field amplitudes, we proposed and demonstrated the possibility to exploit the Stark shift of the hyperfine coupling to accomplish this task. Not only could we tune the level splitting of our nuclear spin qubit, but also the generation of large effective AC magnetic fields at the nucleus was possible. In this way we performed the first electrical manipulation of a single nuclear spin. In combination with the tunability of the resonance frequency by means of a local gate voltage, the addressability of individual nuclear spins in the spirit of Kane's proposal [1] becomes possible.

During my thesis I could demonstrate that single-molecule magnets are potential candidates for quantum bits as defined by DiVincenzo [2] (see Chap. 1 for further explanation):

- **Information storage on qubits**: I could show that information could be stored on the single nuclear spin of an isolated TbPc$_2$ SMM (Chap. 6).
- **Initial state preparation**: the initial state preparation is up to now rather passive, however, due to the quantum nondestructive nature of the measurement scheme, the initial state can be prepared by the measurement itself.
- **Isolation**: since we used a nuclear spin qubit, isolation is one of its intrinsic properties. I could show in Chap. 7 that the dephasing time T_2^* of our nuclear spin qubit is about 64 μs.
- **Gate implementation**: due to a very large effective magnetic field, created by the hyperfine Stark effect, a coherent manipulation of the nuclear spin could be performed within 300 ns, which was 200 times faster than the coherence time T_2.
- **Read-out**: I demonstrated in Chap. 6 that the read-out of the nuclear spin qubit state was performed with fidelities better than 87 % (sample C). Note that this is no intrinsic limitation of the system and will be improved in future experiments.

However, to be competitive with existing qubit systems, the read-out of the nuclear spin state needs to be speed up by at least three orders of magnitude. To achieve this acceleration of the read-out cycle, a different detection scheme is necessary. On of the most established methods in nuclear spin based qubits takes advantage of the

(a)

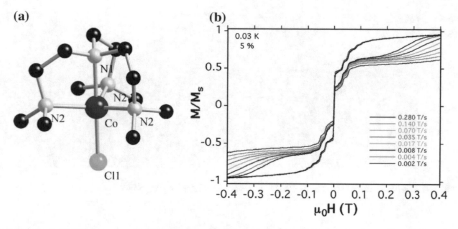

Fig. 8.1 **a** Structure of the Co(Me6tren)Cl SMM. **b** Hysteresis loop of a diluted Co(Me6tren)Cl crystal obtained with a microsquid. Steps in the hysteresis indicate the nuclear spin resolved quantum tunneling of magnetization

nuclear spin resolved electron spin resonance (ESR) as a read-out tool. As already pointed out in Chap. 7, this technique seems to be incompatible TbPc$_2$ SMMs due to the spin ground state of $m_J = \pm 6$ and an excited state separation of about 12.5 THz. However, different kinds of single-molecule magnets with ESR compatible properties could be thought of. Such a molecule is for example the Co(Me6tren)Cl SMM (see Fig. 8.1). It is one of the first mononuclear single-molecule magnets based on transition metal ion. Those molecules are chemically more stable than their polynuclear counterparts, which allows the manipulation and study of its magnetic properties on the single molecule level. It has an Ising type spin ground state of $S = \pm 3/2$, which makes a ESR transition more likely. Furthermore, ^{59}Co is among the 22 existing elements having only one natural isotopic abundance, which is of major importance for its use as a nuclear spin qubit. First, preliminary measurements show, that the hyperfine interaction of the Co^{2+} ion is comparable to the Tb^{3+}, as the steps in the hysteresis curve coming from the hyperfine coupling are well distinguished (see Fig. 8.1(b)). However, its compatibility to the molecular spin-transistor design and the ESR transition between the $S = \pm 3/2$ ground states still needs to be proven. Another important point, which has not been shown yet, is the scalability of our qubits. In contrary to common top-down approaches, like coupling different qubits in a cavity, we want to exploit the potential of organic chemistry in designing molecules including more than one qubit. Starting from the mononuclear terbium double-decker SMM (see Fig. 8.2a), a two qubit system could be made by using a triple-decker SMM with two terbium ions (see Fig. 8.2b). The coupling between the terbium ions is established via the exchange interaction, mediated using an unbound electron of the phthalocyanine ligands. In the single-molecule spin-transistor layout the electron can be easily removed or added by means of the gate voltage. Thus, the coupling between the terbium ions can be switched on or off, allowing for the control of entanglement.

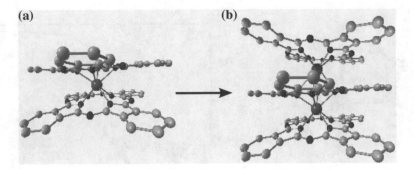

Fig. 8.2 a Structure of the TbPc$_2$ SMM. **b** Structure of the Tb$_2$Pc$_3$ SMM

The results presented in this thesis, extent the potential of molecular spintronics beyond classical data storage. We demonstrated the first experimental evidence of a coherent nuclear spin manipulation inside a single-molecule magnet, and therefore build the foundation for the first molecular quantum bits. Their great versatility holds a lot of promises for a variety of future applications and, maybe one day, a molecular quantum computer.

References

1. B.E. Kane, A silicon-based nuclear spin quantum computer. Nature **393**, 133–137 (1998)
2. D.P. DiVincenzo, *Topics in Quantum Computers*, vol. 345. NATO Advanced Study Institute. Ser. E Appl. Sci. vol. 345 (1996)

Appendix A
Spin

A.1 Charged Particle in a Magnetic Field

If an atom is exposed to an external magnetic field it will experience an interaction which can be quantified by the following Hamiltonian.

$$H = \sum_{i=1}^{Z} \frac{1}{2m_e} \left(p_i + eA(r_i) \right)^2 + V(r_i), \qquad (A.1.1)$$

where p is the momentum operator, $(-e)$ the charge of an electron and A the vector potential. To simplify the calculation, the Coulomb gauge $div A = 0$ can be used, which makes the operators p and A commute. Moreover, the vector potential is chosen to be $A = \frac{1}{2} (B \times r)$. Inserting this into Eq. A.1.1 and expanding the canonical momentum gives:

$$H = \sum_{i=1}^{Z} \frac{p_i^2}{2m_e} + \frac{e}{2m_e} p_i \left(B \times r_i \right) + \frac{e^2}{2m_e} \left(\frac{B \times r_i}{2} \right)^2 + V(r_i) \qquad (A.1.2)$$

Applying the rules for triple products: $p_i (B \times r_i) = B (r_i \times p_i)$, and inserting the electron orbital angular momentum $l_i = r_i \times p_i$ we get:

$$H = \sum_{i=1}^{Z} \frac{p_i^2}{2m_e} + V(r_i) + \frac{e}{2m_e} B \sum_{i=1}^{Z} L_i + \sum_{i=1}^{Z} \frac{e^2 B^2}{8m_e} r_i^2 sin^2(\theta_i) \qquad (A.1.3)$$

Substituting $\sum_{i=1}^{Z} \frac{p_i^2}{2m_e} + V(r_i)$ with the Hamiltonian in absence of a magnetic field H_0, $\frac{e\hbar}{2m_e}$ with the Bohr magneton μ_B and $\sum_{i=1}^{Z} l_i$ with total orbital momentum L, gives rise to final Hamiltonian:

© Springer International Publishing Switzerland 2016
S. Thiele, *Read-Out and Coherent Manipulation of an Isolated Nuclear Spin*,
Springer Theses, DOI 10.1007/978-3-319-24058-9

$$H = H_0 + \underbrace{\mu_B \frac{L}{\hbar} B}_{H_{para}} + \underbrace{\sum_{i=1}^{Z} \frac{e^2 B^2}{8m_e} r_i^2 sin^2(\theta_i)}_{H_{dia}>0} \tag{A.1.4}$$

This Hamiltonian is divided into three parts. The first one, H_0, describes the atom without a magnetic field. In the second term is a scalar product of L and B, which will align two vectors anti-parallel in order to minimize the energy and is responsible for the paramagnetism. The last part of the equation is always positive and therefore increasing the energy of the atom. It describes the diamagnetic response to an applied field. If we put some numbers into the equation, e.g. $|L|/\hbar = \sqrt{2}$, $|B| = 1T$, $r_i^2 = 0.3nm$, we get $H_{para} \approx 100\,\mu eV$ and $H_{dia} \approx 1\,neV$. Thus, the diamagnetism is much smaller than the paramagnetism and is only of significance in systems with closed or half filled shells. The total magnetic moment μ is calculated by taking the first derivative of the Hamilton operator with respect to the magnetic field. Using Eq. A.1.4 results in:

$$\mu = \frac{\partial H}{\partial B} = \underbrace{\mu_B \frac{L}{\hbar}}_{\mu_L} + \underbrace{\sum_{i=1}^{Z} \frac{e^2 B}{4m_e} r_i^2 sin^2(\theta_i)}_{\mu_{ind}}$$

The last term arises only for finite B, and describes the induced magnetic moment. The first term, however, is present also at zero magnetic field and represents a permanent magnetic moment due to the orbital motion:

$$\mu_L = \mu_B \frac{L}{\hbar} \tag{A.1.5}$$

where L is the total orbital angular momentum with its quantum number L. Its modulus is obtained by:

$$|L| = \hbar\sqrt{L(L+1)} \tag{A.1.6}$$

and its projection on the z axis is given as:

$$L_z = \hbar m_L \tag{A.1.7}$$

with m_L being the magnetic orbital quantum number ranging from $-L$ to L, and having therefore $2L + 1$ possible values.

A.2   Electron Spin

When Stern and Gerlach did their famous experiment in 1922, they discovered that electrons posses an internal permanent magnetic moment, which is independent of its orbital motion and takes only two quantized values. In analogy to the orbital angular momentum L it is assumed that an additional intrinsic angular momentum S, which is called spin, gives rise to this permanent magnetic moment. Similar to Eqs. A.1.5–A.1.7 we define a magnetic moment:

$$\mu_S = g_S \mu_B \frac{S}{\hbar} \tag{A.2.1}$$

where g is the Landée factor and S is the total spin with its quantum number S. The modulus of S is calculated as:

$$|S| = \hbar\sqrt{S(S+1)} \tag{A.2.2}$$

and its projection on the z-axis is given by:

$$S_z = \hbar m_S \tag{A.2.3}$$

In contrary to the orbital angular momentum, the magnetic moment is increased by the Landée factor and S can take half integer values.

The Hamiltonian of a charge particle with spin S modifies to:

$$H = H_0 + \mu_B \underbrace{\frac{L + g_S S}{\hbar} B}_{H_{para}} + \underbrace{\sum_{i=1}^{Z} \frac{e^2 B^2}{8 m_e} r_i^2 sin^2(\theta_i)}_{H_{dia} > 0} \tag{A.2.4}$$

A.3   Spin Matrices

Let us first consider a system with only two spin values: $s = {}^1\!/_2$ and $m_S = -{}^1\!/_2, +{}^1\!/_2$. This is a very simple case, but helps understanding more difficult spin system. When calculating energy levels of spin 1/2 systems it is convenient to work with the matrix representation, where the wave function Ψ is a vector with the spin up and down amplitude and the operator S is a vector of two by two matrices $((2S+1) \times (2S+1))$:

$$S = \frac{\hbar}{2} \sigma \tag{A.3.1}$$

where σ are the so-called Pauli matrices:

$$\sigma_x = \begin{pmatrix} 0 & 1 \\ 1 & 0 \end{pmatrix}, \ \sigma_y = \begin{pmatrix} 0 & -i \\ i & 0 \end{pmatrix}, \ \sigma_z = \begin{pmatrix} 1 & 0 \\ 0 & -1 \end{pmatrix}$$

Often, instead of σ_x and σ_y, their linear combinations $\sigma_+ = (\sigma_x + i\sigma_y)$ and $\sigma_- = \sigma_x - i\sigma_y$ are used since they are more adapted to the spin up and spin down basis.

$$\sigma_+ = \begin{pmatrix} 0 & 1 \\ 0 & 0 \end{pmatrix}, \quad \sigma_- = \begin{pmatrix} 0 & 0 \\ 1 & 0 \end{pmatrix} \tag{A.3.2}$$

In this representation the Zeeman energy is calculated by diagonalizing the following Hamiltonian:

$$H_{\text{Zeeman}} = \frac{1}{2}\mu_B gs \left[B_x \begin{pmatrix} 0 & 1 \\ 1 & 0 \end{pmatrix} + B_y \begin{pmatrix} 0 & -i \\ i & 0 \end{pmatrix} + B_z \begin{pmatrix} 1 & 0 \\ 0 & -1 \end{pmatrix} \right]$$

Analogue to the spin 1/2 system those matrices can be calculated for spin systems of order N, where σ_z is a $(2N+1) \times (2N+1)$ matrix with only diagonal elements.

$$\sigma_z(N) = \begin{pmatrix} -N & & 0 \\ & \ddots & \\ 0 & & N \end{pmatrix} \tag{A.3.3}$$

The matrices $\sigma_x(N)$ and $\sigma_y(N)$ are obtained via $\sigma_+(N)$ and $\sigma_-(N)$:

$$\sigma_\pm(N) = \sqrt{N(N+1) - m_N(m_N \pm 1)} \, \delta_{i\pm1,j} \tag{A.3.4}$$

A.4 Dirac Equation and Spin-Orbit Coupling

The origin of spin and therefore spin-orbit interaction lies in the relativistic nature of electrons. Relativity theory teaches us that the energy of an electron is calculated by: $E = \sqrt{c^2 p^2 + m_e^2 c^4}$, where c is the speed of light, m_e the free electron mass and p the relativistic, classical momentum: $p = (1 - v^2/c^2)^{-1/2} m_e v$. Due to its non-linearity it is not so easy to translate this equation using the correspondence principle of quantum mechanics into an operator. The only way to solve this problem is to linearize the above equation. It can be shown, that this is only possible by rewriting the standard representation of the Schrödinger equation in the matrix representation. The idea is to find a matrix which multiplied by itself, gives the energy eigenvalues squared. The solution to this problem was found by Paul Dirac in 1928 and has the following form:

$$H_D = \begin{pmatrix} m_e c^2 & 0 & cp_z & c(p_x - ip_y) \\ 0 & m_e c^2 & c(p_x - ip_y) & -cp_z \\ cp_z & c(p_x - ip_y) & -m_e c^2 & 0 \\ c(p_x - ip_y) & -cp_z & 0 & -m_e c^2 \end{pmatrix}$$

which multiplied by itself gives:

$$H_D^2 = \begin{pmatrix} c^2 p^2 + m_e^2 c^4 & 0 & 0 & 0 \\ 0 & c^2 p^2 + m_e^2 c^4 & 0 & 0 \\ 0 & 0 & c^2 p^2 + m_e^2 c^4 & 0 \\ 0 & 0 & 0 & c^2 p^2 + m_e^2 c^4 \end{pmatrix}$$

The energy eigenvalues of the Dirac Hamiltonian are:

$$E = \pm \sqrt{c^2 p^2 + m_e^2 c^4}$$

Where each eigenvalue is twice degenerate. The positive energies are describing electrons, whereas the negative energies are for positrons. The time independent Dirac equation is then:

$$\begin{pmatrix} m_e c^2 & 0 & cp_z & c(p_x - ip_y) \\ 0 & m_e c^2 & c(p_x - ip_y) & -cp_z \\ cp_z & c(p_x - ip_y) & -m_e c^2 & 0 \\ c(p_x - ip_y) & -cp_z & 0 & -m_e c^2 \end{pmatrix} \begin{pmatrix} \Psi_e^\uparrow \\ \Psi_e^\downarrow \\ \chi_p^\uparrow \\ \chi_p^\downarrow \end{pmatrix} = E \begin{pmatrix} \Psi_e^\uparrow \\ \Psi_e^\downarrow \\ \chi_p^\uparrow \\ \chi_p^\downarrow \end{pmatrix}$$

where $\Psi_e^\uparrow, \Psi_e^\downarrow$ is the up-spin or down spin electron wave function and $\chi_p^\uparrow, \chi_p^\downarrow$ is the up-spin or down-spin positron wave function, respectively. To describe an relativistic electron in an electro-magnetic field the following substitutions are usually made: $p \rightarrow p + eA$ and $E = E + e\phi$. Where A and ϕ are the magnetic vector potential and the electric scalar potential, respectively. In the following we want combine the up-spin and down-spin component to get smaller expressions. It can be shown easily that: $c\boldsymbol{p\sigma} = c(p_x \sigma_x + p_y \sigma_y + p_z \sigma_z) = \begin{pmatrix} cp_z & c(p_x - ip_y) \\ c(p_x - ip_y) & -cp_z \end{pmatrix}$. Thus we get:

$$\begin{pmatrix} m_e c^2 & c(p + eA)\sigma \\ c(p + eA)\sigma & -m_e c^2 \end{pmatrix} \begin{pmatrix} \Psi \\ \phi \end{pmatrix} = (E + e\phi) \begin{pmatrix} \Psi \\ \phi \end{pmatrix}$$

This is a coupled equation of Ψ and ϕ. Expanding this matrix equation results in:

$$\left(E - m_e c^2 + e\phi \right) |\Psi> = c(p + eA)\sigma |\chi> \qquad (A.4.1)$$

$$\left(E + m_e c^2 + e\phi \right) |\chi> = c(p + eA)\sigma |\Psi> \qquad (A.4.2)$$

We can therefore express $|\chi>$ in terms of $|\Psi>$:

$$|\chi> = \frac{c}{\left(E + m_e c^2 + e\phi \right)} (p + eA)\sigma |\Psi>$$

Until now everything is exact. To simplify this equation we use the Taylor series expansion.

$$|\chi> \approx \frac{1}{2m_ec}\left(1-\frac{E-m_ec^2+e\phi}{2m_ec^2}\right)(p+eA)\sigma|\Psi>$$

Substituting this result into Eq. A.4.1 gives:

$$\left(E-m_ec^2+e\phi\right)|\Psi\rangle \approx \frac{1}{2m_e}(p+eA)\sigma\left(1-\frac{E-m_ec^2+e\phi}{2m_ec^2}\right)(p+eA)\sigma|\Psi\rangle$$
$$=\left[\frac{[(p+eA)\sigma]^2}{2m_e}\left(1-\frac{E-m_ec^2}{2m_ec^2}\right)\right.$$
$$\left.-\frac{e}{4m_e^2c^2}(p+eA)\sigma\,(\phi)\,(p+eA)\sigma\right]|\Psi\rangle$$

We used the fact that the operator $(p+eA)\sigma$ is not acting on $E-m_ec^2$. Now we want to expand the second term. To do so we recall that the momentum operator $p=-i\hbar\nabla$ and that $p\phi=\phi p-i\hbar\nabla\phi$. Inserting this into the above equation gives:

$$(p+eA)\sigma\,(\phi)\,(p+eA)\sigma = \phi\left[(p+eA)\sigma\right]^2-i\hbar\left[(\nabla\phi)\,\sigma\right]\left[(p+eA)\,\sigma\right]$$

using the equation: $(X\sigma)\,(Y\sigma)=XY+i\sigma\,(X\times Y)$ we end up with:

$$-i\hbar\,(\nabla\phi)\,\sigma\left[(p+eA)\,\sigma\right]=-i\hbar\,(\nabla\phi)\,(p+eA)+\hbar\sigma\left[(\nabla\phi)\times(p+eA)\right]$$

$$\left(E-m_ec^2+e\phi\right)|\Psi> = \left[\underbrace{\frac{[(p+eA)\sigma]^2}{2m_e}\left(1-\frac{E-m_ec^2+e\phi}{2m_ec^2}\right)}_{\text{Pauli equation + relativistic correction}}|\Psi\rangle\right]$$
$$+\left[\underbrace{\frac{e}{4m_e^2c^2}\nabla\phi\,(p+eA)}_{\text{Darwin−term}}-\underbrace{\frac{\hbar e}{4m_e^2c^2}\sigma\left[\nabla\phi\times(p+eA)\right]}_{\text{spin−orbit−term}}\right]|\psi\rangle$$

We are now concentrating only on the last term, since it is the most interesting for our purposes.

By changing to the spherical coordinate system:

$$\nabla\phi = \frac{1}{r}\frac{d\phi}{dr}r$$

resulting in:

$$\frac{\hbar e}{4m_e^2 c^2} \sigma \left[\nabla \phi \times (p + eA) \right] = \frac{\hbar e}{4m_e^2 c^2} \sigma \left[\frac{1}{r} \frac{d\phi}{dr} r \times (p + eA) \right]$$

Since $p + eA$ is the canonical momentum the expression $r \times (p + eA)$ gives us the orbital momentum l. The term $\frac{1}{r} \frac{d\phi}{dr}$ is just a scalar and can be combined with the pre-factor to the spin-orbit coupling constant $\xi = \frac{\hbar e}{4m_e^2 c^2} \left(\frac{1}{r} \frac{d\phi}{dr} \right)$. Since σ is the operator for the spin s we result in the final one electron spin-orbit Hamiltonian:

$$H_{so} = \xi \, ls$$

If we are now considering systems with more than one electron, there are two possibility of how the spin-orbit coupling effects the orbital energies. The first and for us less interesting case is a system where the spin-orbit coupling is larger than the electron-electron interaction. There each electrons spin s_i couples with its orbit l_i to form an total momentum $j_i = l_i + s_i$. The coupling energy is than given by $H_{l_i, s_i} = c_{ii} l_i s_i$. In the second case the electron-electron interaction, or in other words the coupling between different orbital momenta $H_{l_i l_j} = a_{ij} l_i l_j$ and spins $H_{s_i s_j} = b_{ij} s_i s_j$ is larger than the spin-orbit coupling. Now the different orbital momenta couple to a total orbital momentum $L = \sum_i l_i$ and the different spins couple to a total spin $S = \sum_i s_i$ before coupling the the total momentum $J = L + S$. The spin-orbit coupling energy is than given by: $H_{so} = \lambda \, LS$. With this knowledge we can also try to understand the 3. Hunds rule. Therefore we are relating the one electron spin-orbit coupling constant ξ with λ:

$$H_{so} = \xi \sum_i l_i \sum_i s_i = \lambda LS$$

Therefore

$$\lambda = \frac{\xi \sum_i l_i l_i}{LS}$$

for less than half filled shells s_i is always $\frac{1}{2}$ and can be put in from of the sum. Thus λ becomes positive for less than half filled shells and the ground state is $J = |L - S|$.

$$\lambda = \frac{\frac{1}{2} \xi \sum_i L_i}{LS} = \frac{\xi}{2S} > 0$$

For more than half filled shells s_i has values of $+\frac{1}{2}$ and $-\frac{1}{2}$ and the sum is split in two:

$$\lambda = \frac{\frac{1}{2}\xi \overbrace{\underset{i}{\overset{half}{\sum}} L_i}^{0}}{LS} - \frac{\frac{1}{2}\xi \overbrace{\underset{half}{\overset{n}{\sum}} L_i}^{L}}{LS} = -\frac{\xi}{2S} < 0$$

Now λ becomes negative and $J = L + S$ is the new ground state, since it has the smallest energy.

Appendix B
Stevens Operators

$$O_2^0 = 3J_z^2 - J(J+1)$$

$$O_4^0 = 35J_z^4 - 30J(J+1)J_z^2 + 25J_z^2 - 6J(J+1) + 3J^2(J+1)^2$$

$$O_4^4 = \frac{1}{2}(J_+^4 + J_-^4)$$

$$O_6^0 = 231J_z^6 - 315J(J+1)J_z^4 + 735J_z^4 + 105J^2(J+1)^2J_z^2 - 525J(J+1)J_z^2 +$$
$$+294J_z^2 - 5J^3(J+1)^3 + 40J^2(J+1)^2 - 60J(J+1)$$

$$O_6^4 = \frac{1}{4}\left[(11J_z^2 - J(J+1) - 38)(J_+^4 + J_-^4) + (J_+^4 + J_-^4)(11J_z^2 - J(J+1) - 38)\right]$$

where J_z, J_+ and J_- are the generalized Pauli operators of order N.

© Springer International Publishing Switzerland 2016
S. Thiele, *Read-Out and Coherent Manipulation of an Isolated Nuclear Spin*,
Springer Theses, DOI 10.1007/978-3-319-24058-9

Appendix C
Quantum Monte Carlo Code

The following python code is based on a quantum Monte Carlo algorithm and was used to simulate the nuclear spin trajectory.

```python
from pylab import *
from scipy.optimize.minpack import curve_fit
import pickle
class QMC:
    def __init__(self,T,Gamma,dt,delta_t,p_LZ):
        self.Psi0 = array([0,0,0,1]) # initial state
        self.H0   = array([0.0,121.0,270.0,448.0]) # in mK
        self.dt   = dt # QMC time step
        self.delta_t = delta_t # measurement interval
        self.Gam= Gamma
        self.T    = T
        self.p_LZ = p_LZ
        self.d_omega = diff(self.H0)
        self.n_T = array([1./(exp((self.d_omega[0])/T)-1),
                          1./(exp((self.d_omega[1])/T)-1),
                          1./(exp((self.d_omega[2])/T)-1)])
        self.scl = array([1.0,2.0,2.2])
        self.C1C1 = self.Gam*array([0,
                            self.scl[0]*(1+self.n_T[0]),
                            self.scl[1]*(1+self.n_T[1]),
                            self.scl[2]*(1+self.n_T[2])])
        self.C2C2 = self.Gam*array([self.scl[0]*self.n_T[0],
                            self.scl[1]*self.n_T[1],
                            self.scl[2]*self.n_T[2],
                            0])
        self.H1 = ones(4)-0.5*dt*(self.C1C1+self.C2C2)
```

© Springer International Publishing Switzerland 2016
S. Thiele, *Read-Out and Coherent Manipulation of an Isolated Nuclear Spin*,
Springer Theses, DOI 10.1007/978-3-319-24058-9

```
    self.res  = []
    self.data = []
    self.miss = []
    self.err  = 0


def run(self, steps=2**22, tr_rt=True):

    res  = [] #data are stored every quantum jump
    data = [] #data are stored in discrete time intervals
    miss = []
    Psi  = self.Psi0              #initial state

    t = 0      #lifetime of the current state
    rev = 1    #-1: e-spin down
               #+1: e-spin up

    time = 0 #discrete time
    #number of  cycles for each delta_t
    N = int(delta_t/self.dt)
    #cycle is devided into 5 sections
    t1 = 1.0*N/5.0

    t_arr = round(t1)*ones(10)
    #section numbering
    itv   = array([4,0,1,2,3,4,3,2,1,0])
    #itv[0] current section
    LZ_event = t_arr[0]

    #Monte Carlo Loop
    for ii in range(steps):
        # status report
        if (ii%int(steps/100)==0):
            print str(int(1.*ii/steps*100))+'% completed '

        # increase lifetime of the current state
        t = t + self.dt

        #Thermal contribution
        #-> evolve the population continously
        Psi1 = self.H1*Psi
        dp_rel = dot(Psi1, self.C1C1*Psi1)*self.dt
        dp_exc = dot(Psi1, self.C2C2*Psi1)*self.dt
        dp = dp_rel +dp_exc

        eps = rand()
```

```
      #eps > dp: # nothing happens
   if eps < dp: # quantum jump
        #store population and lifetime every quantum
        #jump for testing
        res.append(Psi[::-rev]*t)
        t = 0 #reset lifetime after quantum jump

        if eps < dp_rel : # relaxation
            Psi = array([Psi[1],Psi[2],Psi[3],0])

        else: #excitation
            Psi = array([0,Psi[0],Psi[1],Psi[2]])


   #Landau Zener contribution
   #every section: possible Landau-Zener transition
   #              if LZT --> inverse and store population
   #Am I at the anticrossing?
   if ((ii == LZ_event) and (tr_rt == True)):
        #cycle through anticrossings
        itv = itv[[1,2,3,4,5,6,7,8,9,0]]
        t_arr = t_arr[[1,2,3,4,5,6,7,8,9,0]]
        # set counter to next anticrossing
        LZ_event += t_arr[0]
        #Am I not at the border?
        if itv[0] < 4:
            #is nuclear spin in the right state?
            if (Psi[::-rev][itv[0]]==1):
                eps_LZ = rand()
                # if rand < Landau-Zener probability
                # -->QTM
                # --> inverse population
                # --> store nuclear spin state

                if (eps_LZ < self.p_LZ):
                    # flip e-spin
                    rev = rev * -1
                    # inverse n-spin population
                    Psi = Psi[::-1]
                    # determine n-spin
                    mj = 1.5-itv[0]
                    #store data
                    data.append([time,mj])
                elif(eps_LZ > self.p_LZ):
                    mj = 1.5-itv[0]
                    miss.append([time,mj])
```

```python
                elif (tr_rt == False):
                    mj = nonzero(Psi==1)[0][0]-1.5   # Psi[0]=1 —> mj=-3/2
                    data.append([time,mj*rev])
            #increase time at the border
            else: time += delta_t

    self.res = array(res)
    self.data = array(data)
    self.miss = array(miss)

def Pop(self,save=False):
    """
    plot histogram of the time average population
    """
    f = plt.figure()
    ax = f.add_subplot(111)
    ax.set_ylabel('pop total');
    ax.set_title('T = '+str(self.T)+' mK');


    p0 = size(nonzero(self.data[:,1]==  1.5))
    p1 = size(nonzero(self.data[:,1]==  0.5))
    p2 = size(nonzero(self.data[:,1]==  -0.5))
    p3 = size(nonzero(self.data[:,1]==  -1.5))
    norm = 1.*(p0+p1+p2+p3)
    ax.bar([-1.5,-0.5,0.5,1.5],array([p0,p1,p2,p3])/norm,
            width=0.4,align='center',alpha=1,
            color=['grey','blue','green','red'])
    ax.set_xticks((-1.5,-0.5,0.5,1.5))
    #ax.invert_xaxis()
    ax.set_xticklabels((r'$|+\frac{3}{2}\rangle$',
                        r'$|+\frac{1}{2}\rangle$',
                        r'$|-\frac{1}{2}\rangle$',
                        r'$|-\frac{3}{2}\rangle$'))

    f.tight_layout()
    if save == True:
        f.savefig('Histogram.png',dpi=300)
        f.savefig('Histogram.pdf')


def DeltaMI(self,save=False):
    temp = []
    for ii in range(size(self.data[:,0])-1):
        temp.append([self.data[ii+1,1]-self.data[ii,1]])
    f1 = plt.figure()
```

```python
        ax1 = f1.add_subplot(111)
        self.dmi = array(temp)
        ax1.hist(self.dmi, bins=(-3,-2,-1,0,1.0,2,3),
                align='left', rwidth=0.5, normed=True, color='r')
        ax1.set_xlabel(r'$\Delta m_{\mathsf{1}}$')
        ax1.set_ylabel('probability')
        ax1.set_xlim((-3,3))
        if save == True:
            f1.savefig('delta_m.png', dpi=300)
            f1.savefig('delta_m.pdf')



    def Lifetime(self, xmax, save = False):
        """
        extract T1
        """
        delta_t = self.delta_t
        temp1 = []
        temp2 = []
        temp3 = []
        temp4 = []

        tau = 0
        for ii in range(size(self.data[:,0])-1):
            if (self.data[ii,1] == self.data[ii+1,1]):
                tau += self.data[ii+1,0]-self.data[ii,0]
            else:
                if (self.data[ii,1] == +1.5): temp1.append(tau)
                if (self.data[ii,1] == +0.5): temp2.append(tau)
                if (self.data[ii,1] == -0.5): temp3.append(tau)
                if (self.data[ii,1] == -1.5): temp4.append(tau)
                tau = 0

        temp = [temp1, temp2, temp3, temp4]

        exp_fit = lambda t, tau, a : a*exp(-t/tau)
        time = linspace(0,120,100)

        f1 = plt.figure(figsize=(12,8))

        for i in range(4):
            ax = f1.add_subplot(2,2,i+1)
            ax.set_yscale('log')
            ax.set_ylim(0.005,1)
            ax.set_yticks((0.01,0.1,1))
```

```python
        ax.set_xlabel(r'$t \ (\mathsf{s})$')
        ax.set_ylabel(r'$\langle m_{\mathsf{l}} \ = \ $'
                      +str(1.5-i)+r"$\rangle $");
        ax.set_xlim((0,xmax));
        ax.set_xticks((0,20,40,60,80,100,120))
        #create Histogramm of lifetime distribution
        H1 = histogram(array(temp[i]),
                       bins=linspace(delta_t,120,
                       int(120/delta_t)))

        #extract all nonzero elements #
        lft = (reshape(concatenate((H1[1][nonzero(H1[0]!=0)],
               H1[0][nonzero(H1[0]!=0)])),
               (size(H1[1][nonzero(H1[0]!=0)]),2),order='F'))

        params, cov = curve_fit(exp_fit,lft[:,0],
                                lft[:,1],[10,100])

        fit = exp(-time/params[0])

        #normalize
        lft[:,1] = lft[:,1]/params[1]
        ax.scatter(lft[:,0],lft[:,1],c='k')
        ax.plot(time,fit,'r—',linewidth=3)
        ax.text(0.6,0.8,r'$\tau$'+' = '
                +str(round(params[0]*100)/100)+'s',
                fontsize='x-large',transform=ax.transAxes)

    f1.tight_layout()
    if save == True:
        f1.savefig('lifetime.png',dpi=300,format='png')
        f1.savefig('lifetime.pdf',dpi=300,format='pdf')

def Write(self, outfile):
    """
    save simulated data
    """
    f = open(outfile, "w+b")
    pickle.dump(self.data, f)
    f.close()


if __name__ == '__main__':

    Gam = 1./41   # 1/s
    T   = 150.0   # temperature in mK
```

```
delta_t = 2.5 # s
dt    = delta_t/60 # s
P_LZ = 0.515  # Landau-Zener probability

sim = QMC(T,Gam,dt,delta_t,P_LZ)
sim.run(2**24)
sim.Pop(True)
sim.DeltaMI(True)
sim.Lifetime(120,True)
#sim.Write("sim_data.file")
```

Appendix D
Qutip Code

The following python program was used to simulate the trajectory of the Bloch vector.

```python
from qutip import *
from pylab import *
from numpy import real
from mpl_toolkits.mplot3d import Axes3D
import mpl_toolkits.mplot3d.axes3d as p3
def run():
    #
    # problem parameters:
    #
    delta = 0 * 2 * pi      # qubit sigma_x coefficient
    omega = 1.0 * 2 * pi    # qubit sigma_z coefficient
    A = 0.25 * 2 * pi       # driving amplitude
    w = 1.0 * 2 * pi        # driving frequency
    gamma1 = 0.0            # relaxation rate
    n_th = 0.0              # average number of excitations
    psi0 = basis(2, 0)      # initial state
    #
    # Hamiltonian
    #
    sx = sigmax(); sy = sigmay(); sz = sigmaz();
    sm = destroy(2);
    H0 = - (delta + omega) / 2.0 * sz
    H1 = - A * sx
    #
    # define the time-dependence of the Hamiltonian
    #
    args = {'w': w}
    Ht = [H0, [H1, 'sin(w*t)']]
```

© Springer International Publishing Switzerland 2016
S. Thiele, *Read-Out and Coherent Manipulation of an Isolated Nuclear Spin*,
Springer Theses, DOI 10.1007/978-3-319-24058-9

```
#
# collapse operators
#
c_op_list = []

rate = gamma1 * (1 + n_th)
if rate > 0.0:
    c_op_list.append(sqrt(rate) * sm)          # relaxation

rate = gamma1 * n_th
if rate > 0.0:
    c_op_list.append(sqrt(rate) * sm.dag())  # excitation


#
# evolve and system subject to the time-dependent hamiltonian
#
tlist = linspace(0, 0.70 * pi / A, 100)
output1x = mesolve(Ht, psi0, tlist, c_op_list, [sx], args)
output1y = mesolve(Ht, psi0, tlist, c_op_list, [sy], args)
output1z = mesolve(Ht, psi0, tlist, c_op_list, [sz], args)


#
# Alternative: write the Hamiltonian in a rotating frame,
# and neglect the high frequency component (RWA) so that
# the resulting Hamiltonian is time-independent.
#
H_rwa = - delta / 2.0 * sz - A * sx / 2
output2x = mesolve(H_rwa, psi0, tlist, c_op_list, [sx])
output2y = mesolve(H_rwa, psi0, tlist, c_op_list, [sy])
output2z = mesolve(H_rwa, psi0, tlist, c_op_list, [sz])


#
# Plot the solution
#
fig = figure(figsize=(14,7))
rec1 = [0,0,0.5,1]; rec2 = [0.5,0,0.5,1]
rec3 = [0.,0.9,1,0.1]
ax = Axes3D(fig, rec1, azim=-60, elev=30)
ax2 = Axes3D(fig, rec2, azim=-60, elev=30)
ax3 = fig.add_axes(rec3)
ax3.axis("off")
ax3.text(0.05,0,r"$(\sf{a})$",fontsize=35)
ax3.text(0.55,0,r"$(\sf{b})$",fontsize=35)

b1 = Bloch(fig, ax)
```

```python
        b1.add_text(0,0,1.2,"(a)",100)
        b1.font_size = 35; b1.zlabel = [r'$z$','']
        b1.xlpos = [1.3,-1.3];  b1.ylpos = [1.2,-1.2]
        b1.zlpos = [1.2,-1.2]
        b1.vector_color=([0.5,0.5,0.5],[0,0,0])
        b1.vector_mutation = 20
        b1.add_vectors([0,0,1])
        b1.add_vectors([real(output1x.expect[0])[-1],
                        real(output1y.expect[0])[-1],
                        real(output1z.expect[0])[-1]])
        b1.point_color = ("blue")
        b1.add_points([real(output1x.expect[0]),
                       real(output1y.expect[0]),
                       real(output1z.expect[0])])
        b1.draw()
        b2 = Bloch(fig,ax2)
        b2.font_size = 35
        b2.zlabel = [r'$z$','']; b2.xlpos = [1.3,-1.3];
        b2.ylpos = [1.2,-1.2]; b2.zlpos = [1.2,-1.2];

        b2.vector_color=([0.5,0.5,0.5],[0,0,0])
        b2.vector_mutation = 20
        b2.add_vectors([0,0,1])
        b2.add_vectors([real(output2x.expect[0])[-1],
                        real(output2y.expect[0])[-1],
                        real(output2z.expect[0])[-1]])
        b2.point_color = ("blue")
        b2.add_points([real(output2x.expect[0]),
                       real(output2y.expect[0]),
                       real(output2z.expect[0])])
        b2.draw()
        return output1x

if __name__ == '__main__':

    out = run()
```

Curriculum Vitae

Dr. Stefan Thiele

Physicist

Date of Birth 25 Feb 1986	**Contact Information** Address: Laubisruetistrasse 50 8712 Staefa
Nationality German	Email: stefan.thiele@sensirion.com
Current Affiliation Sensirion AG (Switzerland)	Phone: 0041 / 44 300 40 22

[research interest]

- solid state sensors
- MEMs devices
- quantum physics
- spintronics
- graphene based devices
- semiconductor physics

[education]

2010 – 2014 **PhD Thesis:** Nano Physics   **CNRS / Grenoble / France**
Title: *"Read-out and coherent manipulation of an isolated nuclear spin using single-molecule magnet spin-transistor"*
focus on: - nano-physics
- molecular magnets & spintronics
- quantum electronics & quantum computation
- low-temperature physics & cryogenics

2004 – 2010 **Master Thesis:** Applied Physics   **TU Ilmenau / Germany**
MIT / Cambridge / USA
Title: *"Modeling of the AC and DC characteristics of large-area graphene field-effect transistors"*
major in: - semiconductor/ micro and nano-electronics
- laser physics and measurement technology
- solid state and surface physics
minor in: - electronics, mechanics, business studies

1996 – 2004 **High school graduation**   **Zeulenroda / Germany**
graduated with distinction (1.1)

[awards]

2015 Springer thesis award

Thesis award from the Fondation NanoScience

2010 TU Ilmenau - award for being one of the best graduates of the year

2007 – 2009 Scholarship from the German national academic foundation (available to the best 0.5% of all students)

2004 High school – award for being the best graduate of the year

[publications]

2014	*Science* **344**, 6188	2010	*IEEE Trans. on electron Devices* **57**, 3231
2013	*Physical Review Letters* **111**, 037203		*J. of Applied Physics* **107**, *094505*
2012	*Physical Review Letters* **109**, 264301		*Applied Physics Letters* **96**, *123506*
			Nanotechnology **21**, *015601*
2011	*J. of Applied Physics* **110**, *034506*	2009	*Nano Research* **2**, 509

[Web]

 LinkedIn.com

© Springer International Publishing Switzerland 2016
S. Thiele, *Read-Out and Coherent Manipulation of an Isolated Nuclear Spin*,
Springer Theses, DOI 10.1007/978-3-319-24058-9

Printed in the United States
By Bookmasters